Xiao Liu

Bioinformatic view of kinesin super-family and structure of kinesin-1

AF061186

Xiao Liu

Bioinformatic view of kinesin super-family and structure of kinesin-1

large scale analysis of kinesin super-family and structural changes from a closed to an open state of the motor domain of kinesin-1

Südwestdeutscher Verlag für Hochschulschriften

Impressum/Imprint (nur für Deutschland/ only for Germany)
Bibliografische Information der Deutschen Nationalbibliothek: Die Deutsche Nationalbibliothek verzeichnet diese Publikation in der Deutschen Nationalbibliografie; detaillierte bibliografische Daten sind im Internet über http://dnb.d-nb.de abrufbar.

Alle in diesem Buch genannten Marken und Produktnamen unterliegen warenzeichen-, marken- oder patentrechtlichem Schutz bzw. sind Warenzeichen oder eingetragene Warenzeichen der jeweiligen Inhaber. Die Wiedergabe von Marken, Produktnamen, Gebrauchsnamen, Handelsnamen, Warenbezeichnungen u.s.w. in diesem Werk berechtigt auch ohne besondere Kennzeichnung nicht zu der Annahme, dass solche Namen im Sinne der Warenzeichen- und Markenschutzgesetzgebung als frei zu betrachten wären und daher von jedermann benutzt werden dürften.

Verlag: Südwestdeutscher Verlag für Hochschulschriften Aktiengesellschaft & Co. KG
Dudweiler Landstr. 99, 66123 Saarbrücken, Deutschland
Telefon +49 681 37 20 271-1, Telefax +49 681 37 20 271-0
Email: info@svh-verlag.de
Zugl.: Muenchen, LMU, Diss., 2009

Herstellung in Deutschland:
Schaltungsdienst Lange o.H.G., Berlin
Books on Demand GmbH, Norderstedt
Reha GmbH, Saarbrücken
Amazon Distribution GmbH, Leipzig
ISBN: 978-3-8381-1377-7

Imprint (only for USA, GB)
Bibliographic information published by the Deutsche Nationalbibliothek: The Deutsche Nationalbibliothek lists this publication in the Deutsche Nationalbibliografie; detailed bibliographic data are available in the Internet at http://dnb.d-nb.de.

Any brand names and product names mentioned in this book are subject to trademark, brand or patent protection and are trademarks or registered trademarks of their respective holders. The use of brand names, product names, common names, trade names, product descriptions etc. even without a particular marking in this works is in no way to be construed to mean that such names may be regarded as unrestricted in respect of trademark and brand protection legislation and could thus be used by anyone.

Publisher: Südwestdeutscher Verlag für Hochschulschriften Aktiengesellschaft & Co. KG
Dudweiler Landstr. 99, 66123 Saarbrücken, Germany
Phone +49 681 37 20 271-1, Fax +49 681 37 20 271-0
Email: info@svh-verlag.de

Printed in the U.S.A.
Printed in the U.K. by (see last page)
ISBN: 978-3-8381-1377-7

Copyright © 2010 by the author and Südwestdeutscher Verlag für Hochschulschriften Aktiengesellschaft & Co. KG and licensors
All rights reserved. Saarbrücken 2010

Summary

Kinesins form a large microtubule-associated motor protein super-family that can be found in every eukaryotic genome sequenced so far. Not only is the translocation of a large number of organelles, protein complexes and mRNAs carried out by them, but also the formation of the meiotic spindle and mitotic spindle integrity are strongly dependent on the kinesins.

Fourteen different sub-families of kinesin have been reported. However, previous analyses were based on a relatively small number of selected kinesins (<600 sequences). Whether new classes of kinesin exist or the old classification system will hold as new sequence data become available is unknown.

In this project, comprehensive computational analyses were performed on a large kinesin dataset (2,530 sequences). Sixteen conserved motifs were identified within the motor domain, including the ATP-binding motifs, microtubule-binding interface and many conserved secondary structural elements. Phylogenetic analysis confirmed the fourteen sub-family classification scheme. Thirteen of sub-families were well defined and statistically supported. The kinesin-12 sub-family had less support, with a clade confidence of 73%.

In addition, a profile-based, automatic classification program was implemented according to the fourteen kinesin sub-groups. The accuracy of the program is over 85%, which makes the detection and classification of new kinesin sequences fast and easy.

Kinesin-1, formerly known as conventional kinesin, is the best-studied member of the kinesin super-family. Motility studies have revealed an interesting phenomenon that the fungal kinesin-1s move 4-5 times faster than the animal kinesin-1s in general. Determining the sequence and structural factors that are responsible for the velocity difference is a topic of current research. Previous protein-chimera experiments have determined that the motor domain is essential for speed control. However, detailed analyses of the motor domain through mutagenesis have presented many challenges to biologists, because it is still unknown whether the speed is controlled by one particular amino acid residue or by a complex combination of several residues.

With comparative analyses of the primary sequences of fungal and animal kinesin-1s, many group-specific residues were identified. Several of them are located inside functionally important motifs such as the ATP-binding pocket and potential microtubule binding motifs, which appear to be responsible for the functional differences. The others are widely distributed in many important secondary structural elements.

The mapping of these residues onto the fungal and animal three-dimensional crystal structures (1BG2 and 1GOJ) has led to the discovery of several structural changes from a closed to an open conformation of the motor domain. Most of the group-specific residues are involved in the spatial interactions with other group-specific residues or conserved residues. Many of these interactions can be detected only in the closed conformation. They contain functional elements, such as the switch-I, loop-11, β5 etc that lie within the core structure of the motor domain. When the structure changes into the open conformation, these elements are released and become active for binding to the microtubule. At the same time, many new interactions made by the group-specific residues are formed for the stabilization of the motor domain.

These structurally crucial interaction-pairs of residues and the group specific residues found in the ATP-binding pocket provide insight into the potential control of kinesin velocity. The different structures of the fungal and animal ATP-binding pockets appear to be vital for ATP hydrolysis, but cannot control the velocity by itself. Some of the detected combinations of residues must interact within the ATP-binding pocket. They could be used as guidance for the biologists to design experiments to eventually discover the mechanism of velocity control.

Methods developed in this work have proven to be useful for analyzing the kinesins. Analyses of kinesin-1 are only the first step to understand the kinesin super-family. These methods can be applied to other kinesin sub-families. On the other hand, the number of kinesin sequences in public databases is increasing rapidly. In this project, a kinesin web-server has been developed to assist with further research of the kinesins. It stores and classifies all currently identified kinesins and is automatically updated to keep all kinesin ssequences up-to-date. Many useful methods are implemented in the web-server, such as a classification tool, a conservation calculation tool, a motif search tool, and a discriminating residues (group-specific residues) search tool. The user can use these tools to analyze pre-defined kinesin sub-families or user-defined sequences or alignments. In addition, the group-specific residues can be mapped onto user selected

3D structures for direct visual comparison. The web-server is accessible at http://www.bio.uni-muenchen.de/~liu/kinesin_new/

1 Introduction 1
  1.1 Biological background 1
    1.1.1 Motor proteins 1
      1.1.1.1 Dynein 1
      1.1.1.2 Myosin 2
      1.1.1.3 Kinesin 3
  1.2 Bioinformatics and protein research 5
    1.2.1 Sequence databases 6
    1.2.2 Homology search via Basic Local Alignment Search Tool (BLAST) 6
    1.2.3 Comparative proteomics 8
    1.2.4 Phylogeny estimations 8
    1.2.5 Reconstruction of the ancestral state of a protein 11
    1.2.6 Investigation of the evolutionary pathway 13
    1.2.7 Structure comparison 13
  1.3 kinesin-1 project 14
    1.3.1 kinesin super-family 14
    1.3.2 kinesin-1 sub-family 16
    1.3.3 Dichotomy of kinesin-1 in phylogeny and in motility 17
    1.3.4 Previous laboratory research on motility 18
    1.3.5 Attempts by Bioinformatics 21
      1.3.5.1 Comparative approaches 21
      1.3.5.2 Resurrecting ancestral kinesin-1 proteins 23
2 Materials and methods 27
  2.1 Data collection 27
    2.1.1 NCBI RefSeq database 27
    2.1.2 Kinesins in the RefSeq database 28

    2.1.3   PSI-BLAST search against the RefSeq database 29

  2.2   Classification of kinesin sequences 30

    2.2.1   NCBI CDD classification method 30

    2.2.2   Hidden Markov model classification 31

    2.2.3   Significance test 31

        2.2.3.1   E-value threshold cutoff 31

        2.2.3.2   Likelihood ratio test 32

    2.2.4   Automatic classification of kinesin sequences 33

    2.2.5   Nomenclature of kinesins 34

  2.3   Automatic update of the kinesin database 34

  2.4   Multiple sequence alignments 35

  2.5   Comparative proteomics 35

    2.5.1   Conserved amino acids in kinesin sequences 36

    2.5.2   Identifying conserved sequence motifs 38

    2.5.3   Finding known motifs in proteins 40

    2.5.4   Discriminating amino acids 41

  2.6   Phylogeny reconstruction 41

    2.6.1   Methods 41

    2.6.2   Visualization 42

  2.7   Reconstruction of ancestral kinesins 42

  2.8   Structural comparison 42

  2.9   Web server 43

    2.9.1   MySQL database 43

    2.9.2   Web interface 44

3   Results 47

  3.1   Statistics of kinesin sequence data 47

    3.1.1   Kinesin sequences in the RefSeq database 47

    3.1.2   Organisms which contain kinesins 50

3.2 Conservation analyses 53
   3.2.1 Common conserved residues in the kinesin super-family 53
   3.2.2 Common motifs of kinesins 59
      3.2.2.1 ATP-binding pocket 60
      3.2.2.2 Switch I and switch II motifs 61
3.3 Phylogenetic analysis and kinesin classification 62
   3.3.1 Classification based on the Conserved Domain Database 63
   3.3.2 Phylogenetic tree of the kinesin super family 63
   3.3.3 Evolution of kinesins 66
3.4 Kinesin-1 68
   3.4.1 Sequence data 68
   3.4.2 Conservation of kinesin-1 70
   3.4.3 Motif structure of the kinesin-1 sub-family 74
   3.4.4 Phylogeny of kinesin-1 77
   3.4.5 Fungal vs animal kinesin-1 78
   3.4.6 Ancestral kinesin-1 prediction 80
   3.4.7 Discriminating positions in fungal and animal kinesin-1 84
   3.4.8 Discriminating residues and the motor domain structure 87
      3.4.8.1 The bond between 204E and 257K in NcK1 88
      3.4.8.2 Residue 203Q in switch I of NcK1 89
      3.4.8.3 Residues 91, 92, 95, 96 in the ATP-binding pocket of NcK1 90
      3.4.8.4 Residue 243 in NcK1 91
      3.4.8.5 Microtubule binding sites 92
      3.4.8.6 Residues I130 and V173 in the HsK1 95
      3.4.8.7 Residues K226 and D288 in the HsK1 95
      3.4.8.8 Residues S266 and R321 in the HsK1 97
      3.4.8.9 V148 and D144 in the HsK1 97

3.4.9 Putative cooperative residues combinations 99
    3.4.9.1 The second layer of the ATP-binding pocket 102
3.4.10 Conjecture about the velocity controller 103
    3.4.10.1 ATP-binding pocket could be one candidate 103
    3.4.10.2 Other potential factors for the velocity difference 104

4 Conclusion and discussion 107
4.1 Confidence evaluation of the phylogenetic tree 107
4.2 The 14th kinesin sub-family 108
4.3 Bioinformatic approaches to study on kinesin-1 velocity 110
4.4 Outlook 111

5 Appendix 113
5.1 Web-server 113
  5.1.1 Classification of user defined sequence 113
  5.1.2 Conservation calculation tool 114
  5.1.3 Motif search 115
  5.1.4 Pattern search for known motifs 116
  5.1.5 Discriminating residues search tool 117
  5.1.6 Displaying discriminating residues in 3D structures 117
  5.1.7 Basic information of the kinesins 118

6 Bibliography 123
6.1 Publications 128

1 Introduction

1.1 Biological background

1.1.1 Motor proteins

Motor proteins make up a large protein family of ATPases. This family contains some of the most important proteins required for life and is essential for most eukaryotic organisms. As the name implies, motor proteins work like motors, generating mechanical energy from chemical energy released by the hydrolysis of ATP to power movement. Generation of force for muscle contractions, transport of different organelles along microtubules within a cell, and generation of energy for mitosis and meiosis are some examples of essential functions that require motor proteins [1].

There are three groups of motor proteins, myosins, dyneins and kinesins. Myosins are actin associated motor proteins, while dyneins and kinesins are microtubule associated. The biochemical and mechanical properties of motor proteins can be measured precisely. The major functions and structure of these motor proteins were discovered after decades of intensive research. However, many other fundamental questions still remain unanswered. For example, how is the chemical energy transformed into mechanical force? What is the relationship between motility and structure?

With the help of bioinformatics and the continual sequencing of genes, we can better investigate the evolution of the structure and function of each motor protein and eventually address the mechanism of motion of motor proteins.

1.1.1.1 Dynein

Dynein is a microtubule associated motor protein. It is a large molecular protein complex, which has a mass of over one megadalton and consists of 9 to 12 polypeptide chains. Most of the polypeptide chains are common components, although some are unique subunits to specialized dyneins. Depending on the location of action, dynein is classified into two major groups, cytoplasmic dynein and axonemal dynein. Axonemal dynein was first discovered in 1963 and is responsible for movement of cilia and flagella. Cytoplasmic dynein was isolated and identified 20 years later and is essential for the positioning and transportation of various organelles needed for cellular function [2].

Axonemal dynein can be found only in cells that have axonemes of cilia and flagella, while cytoplasmic dynein is expressed in almost all animal cells. It is involved in organelle transport, centrosome assembly and mitosis. [2].

Cytoplasmic dynein is thought to help the Golgi apparatus to position and transport other vesicles made by the endoplasmic reticulum, endosomes and lysosomes to various destinations in the cell in order to perform cellular functions. Cytoplasmic dynein is also crucial in the movement of chromosomes and positioning of mitotic spindles for cell division [2].

In vitro experiments indicate that dynein is a minus-end directed motor, which means it transports cargo along microtubules towards the minus end of the microtubule to the cell center.

1.1.1.2 Myosin

Myosin is a large protein family, and many divergent myosin genes have been found throughout eukaryotic phyla. Alone in human, there are more than 40 different myosin genes.

The term myosin was originally used to describe a group of similar, but non-identical ATPases found in striated and smooth muscle cells [3]. Unlike dynein, myosin is an actin associated motor protein. Myosins are mainly responsible for muscle contractions.

Currently, 18 different classes of myosin are known [4]; however, other researchers claim that there are over 30 different classes [5]. Despite the differences in nomenclature, myosins share highly similar structures; most contain a head, a neck and a tail domain. The head domain of myosin is highly conserved, while the tail region is rather divergent. This phenomenon can be explained by the different functions of the domains. The head domain is essential for actin binding, force generation and movement along actin. Its role in these common functions is thought to be the main reason why the structure of the head domain is so well-conserved. On the other hand, the tail is responsible for cargo binding. Variability in the tail domain is proportional to the number of dissimilar cargoes.

Myosin is involved in many major cellular functions. For example, Myosin II is responsible for muscle contraction, myosin I, IV and V function in vesicle transport and myosin VII is required for spermatogenesis or stereocilia formation [6]. However, the functions of most myosins, as well as their structures, remain unknown.

1.1.1.3 Kinesin

In the mid 1980s, scientists discovered the existence of a group of proteins that can hydrolyze ATP and function as transporters in cells. One of these proteins was dynein. The existence of other motor proteins was unknown at that time. Lasek and Brady (1985) published an article entitled 'Attachment of transported vesicles to microtubules in axoplasm is facilitated by AMP-PNP' in Nature [7]. They reported that AMP-PNP, a non-hydrolysable analogue of ATP, can inhibit vesicle transport in axoplasm. Relatively stable complexes were formed by vesicles and microtubules, which indicate significantly different enzymatic machinery in the dynein-microtubule system. This motor protein was then partially purified from axoplasm in the squid giant axon by Vale and Reese soon after in the same year [8]. They found that this soluble protein induces movement of microtubules on glass, latex beads on microtubules and axoplasmic organelles on microtubules. The protein had an apparent molecular weight of 600 kilodaltons and contained 110-120 and 60-70 kilodalton polypeptides, which were distinct in both molecular weight and enzymatic behavior from dynein and myosin. Therefore, they claimed that a novel class of force generating molecules was found and named them kinesin [8].

Kinesin is the third class of motor proteins that was isolated and is also the first large protein family to be identified in mammals. Kinesins are key players in the intracellular transport system, which is essential for cellular function and morphology. The most important functions of kinesin have been uncovered through numerous molecular biological and genetic approaches during the last few decades.

Kinesins are critical for cellular morphogenesis, functioning and survival. They transport various organelles like mitochondria and Golgi apparati, as well as protein complexes and mRNAs. Recent research also indicates that they are involved in different fundamental processes of life such as brain wiring, memory, learning, activity-dependent neuronal survival during brain development, left-right asymmetry formation, and suppression of tumorigenesis [9].

Similar to dynein, kinesin is also a microtubule associated ATPase. That is, it binds to the microtubule and converts chemical energy released by hydrolysis of ATP to mechanical force in order to walk along the microtubule. In contrast to dynein, the movement on the microtubule is plus-end directed and transports cargo from the cell center to periphery. A known exception is members of the kinesin-14 sub-family, which move toward the minus-end like dynein [10].

Kinesins typically have a common structure that is characterized by a dimer with a motor domain, coiled coil stalk and light chain. Similar to the motor domain in dynein and myosin, the head domain is responsible for microtubule binding [11], ATP binding and hydrolysis [12]. The sequence of the head domain shows a high level of conservation while the tail domain shows great variability due to the diversity of cargo. The location of the motor domain varies among kinesins. For example, kinesin-1 has an N-terminal motor domain, while kinesin-14 has a C-terminal motor domain. One-headed kinesins also exist. An example is KIF1A, which shows similar motility properties to other two-headed kinesins [13].

The kinesin super-family of molecular motors can be subdivided into 14 sub-families based on sequence features and cellular function [14]. Miki (2005) generated a kinesin family tree using about 600 sequences and supported the 14 sub-family classification systems of kinesins [16], while Wickstead and Gull claimed that there were new kinesin sub-families in their 'holistic' kinesin phylogeny [17]. With the growth of sequence databases, the number of available kinesin sequences is expanding rapidly. It is now possible to recreate a reliable phylogenetic tree for the kinesin super-family with a vast number of sequences and settle issues related to kinesin sub-family classification.

Although we know the basic functions and structure of kinesins, many details are yet to be understood. For example, studies have revealed the location of ATP binding sites but not how ATP hydrolysis causes conformational changes. It is also known that kinesins can move progressively along microtubules, but we still do not know the details of this progress. Kinesins transport various cargoes, but we do not know how the tail domain binds its cargo. Finally, we know that some kinesins move fast while some move slowly, but we do not know what aspects of the sequence or structure are responsible for this phenomenon.

1.2 Bioinformatics and protein research

The volume of biological data has grown exponentially within the last few decades, but classical molecular biology methods are far too slow to analyze most of the data. Since the discovery of an infection barrier in *E.coli* K-12 in 1953, there has been over 50 years of molecular studies by numerous research groups all over the world, and only half of its proteins have been experimentally investigated. In 2001, there were only 785,143 proteins from 2005 taxa available in public databases and within eight years time there are over 6,413,124 proteins from 7,773 taxa stored in today's databases [18]. Because classical research methods were no longer efficient, bioinformatics was born as a solution to this problem, by applying computer science and information technology to molecular biology, creating databases for maintaining biological data, developing algorithms and theories to accelerate the speed of protein research and finally combining mathematical approaches and statistical models to gain understanding of biological processes. In the last few decades, bioinformaticians have developed many algorithms and applications to analyze and interpret biological data and to assess their relationships. Protein research has become much easier since the ontology of bioinformatics in finding gene locations, searching for homology, predicting structures and functions, and clustering protein sequences into families.

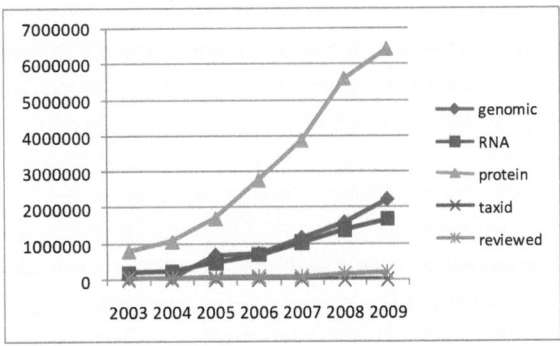

Figure 1. Increase in number of biological entries in the NCBI RefSeq database from release 1 published in 2003 to release 33 published in January,.2009. The number of proteins increases exponentially, while reviewed protein changes increase almost linearly.

1.2.1 Sequence databases

With the ever-increasing production of genomic data, such as DNA sequences and amino acid sequences, creation and maintenance of biological databases has become essential for scientists to access existing data as well as submit new data and revise data. Much of a bioinformaticians' work is concerned with databases. These databases can be public like GenBank [20] or Protein Data Bank (PDB) [21], or private databases that are created by research groups.

A few popular databases are GenBank from the National Center for Biotechnology Information (NCBI) [22], SwissProt from the Swiss Institute of Bioinformatics [23], the Protein information resource (PIR) [24] and the European Molecular Biology Laboratory (EMBL) nucleotide sequence database [25]. NCBI and EMBL are two major nucleotide databases, which collaborate and synchronize their data. These databases are updated on a daily basis, but still many mistakes can be found in the databases due to a high rate of increase in data volume. Common errors are duplicated sequences, sequencing errors and missing or incorrect annotations. In order for researchers to use data from a database, data quality has to be of top priority. Data need to be examined either in the database itself or by cross checking the experimental references available in other databases.

1.2.2 Homology search via Basic Local Alignment Search Tool (BLAST)

Sequence homology refers to sequence similarity due to evolution from a common ancestor. An interspecific or intraspecific gene comparison can show similarities in protein functions or relations between species. There are two types of homology: orthology and paralogy. Sequences are orthologous if they were separated by a speciation event and paralogous if they are separated by a gene duplication event.

Computer programs such as BLAST [26] scan large databases with incredible speed and high sensitivity. Before BLAST was developed, database scanning was very time consuming. BLAST uses a heuristic approach that approximates the Smith-Waterman algorithm to match subsequences in the database to subsequences in the query.

The algorithm of BLAST is described briefly as follows [27]:

1. Remove low complexity regions or sequence repeats in the query sequence.
2. Make a k-letter word list of the query sequence
3. List the possible matching words
4. Organize the remaining high-scoring words into an efficient search tree
5. Repeats step 1 to 4 for each 3-letter word in the query sequence
6. Scan the database sequences for an exact match with the remaining high-scoring words
7. Extend the exact matches to high-scoring segment pairs (HSPs)
8. List all of the HSPs in the database whose score is higher than a given cutoff
9. Evaluate the significance of the HSP score
10. Combine two or more HSP regions into a longer alignment
11. Show the gapped Smith-Waterman local alignments of the query and each of the matched database sequences
12. Report matches whose expected score is lower than a threshold parameter e-value.

BLAST is a software package including many programs, each of them used for a specific type of biological data. For example, blastp searches protein sequences against protein databases, while blastn searches nucleotide sequences against nucleotide databases. A full list of all programs is shown in table 1

Nucleotide blast	**Search a nucleotide database using a nucleotide query** *Algorithms:* **blastn, megablast, discontiguous megablast**
Protein blast	**Search protein database using a protein query** *Algorithms:* blastp, psi-blast, phi-blast
Blastx	Search protein database using a translated nucleotide query

Tblastn	Search translated nucleotide database using a protein query
tblastx	Search translated nucleotide database using a translated nucleotide query

Table 1. Different BLAST applications and their usages [27]

1.2.3 Comparative proteomics

As mentioned before, homologous proteins contain valuable information about processes of protein evolution. Functional and structural properties of proteins can be revealed by comparing sequences and structures of homologous proteins.

Comparative proteomics is an important approach for today's protein research. It is widely used to predict functions and structures for unknown proteins, detect positive selection, and even design drugs.

A common approach of comparative proteomics is to align homologous proteins from different species, and then use different computational methods to address various questions, such as: What type of selection affects the protein's evolution? Which region(s) of the sequence is important for protein function or structure? What are the differences among homologous sequences and how do they relate to protein functions?

1.2.4 Phylogeny estimations

Phylogenetics has been an important research field since Darwin presented his theory of evolution. It is the study of finding the origin of all living beings and the relationship among species. Classical phylogenetics uses the phenotype to classify species, for example size, color, number of legs, wings, etc. It also includes biological and biochemical properties. However, using the phenotype is limited because through convergent evolution, two species of different lineages may have evolved the same phenotype, rendering the two species indistinguishable using phenetics.

With the development of molecular biology, it is known that protein and nucleotide sequences have evolved from a common ancestor over a long period of time. Most of the evolutionary events occurring in these molecules were encoded in their primary sequence. Therefore, phylogenetics at the molecular level is more reliable, because it uses information encoded directly in protein sequences as properties of a species. The basic theory of molecular phylogeny is based on evolution; the number of changes among sequences is positively correlated with the time since their divergence from a common ancestor. In other words, distantly related organisms show greater dissimilarity in sequences, while more closely related organisms show a greater degree of similarity.

In general, phylogenic reconstruction is based on the molecular clock hypothesis, which states that the evolutionary rate of a biological molecule is constant over time. This hypothesis was first proposed by Emile Zuckerkandl and Linus Pauling in 1962, who estimated from fossil evidence that the number of amino acid differences in hemoglobin between different lineages roughly correlated with their divergence [28]. Later in 1963, the phenomenon of genetic equidistance was noted by comparing the number of amino acid differences among cytochrome C of several species [29]. The discovery of the molecular clock has provided a powerful way to time the processes of molecular evolution. Information obtained from molecular genetics can be used in the formation of phylogenetic trees, establishing the dates of evolutionary events such as gene duplication, and discovering the divergence times of genes or taxa. However, the reliability of the molecular clock can be limited by many factors such as generation time, population size, species-specific differences, positive selection, etc. These limitations should be considered, especially when studying long timescales [30].

A classical phylogenetic tree is an acyclic connected graph consisting of a set of linked nodes that represent the evolutionary relationship of a set of biological units, such as DNA sequences, protein sequences, or species. The external nodes of the tree represent species or genes/proteins. Internal nodes represent the hypothetical most recent common ancestor of its descendants and show the time point when an evolutionary event like speciation or gene duplication occurred. Branches between nodes indicate the relationship of linked biological data by using the branch length to represent the number of changes in a molecular sequence a lineage has acquired.

A phylogenetic tree can be rooted or unrooted. When rooted, the root of the tree represents the common ancestor; evolutionary pathways can be estimated from this kind of tree. The path between any node and the root should indicate the evolutionary time or evolutionary distance. An unrooted tree shows only the relationship between analyzed sequence/species without information about the order of branching events. Unrooted trees can be transformed into rooted tree by using an out group that is more distantly related to all other nodes in the tree than those nodes are to each other. For example, the orangutan can be used as out-group for human and chimpanzee, placing the root of the tree on the branch between orangutan and the common ancestor of human and chimpanzee (Figure 2).

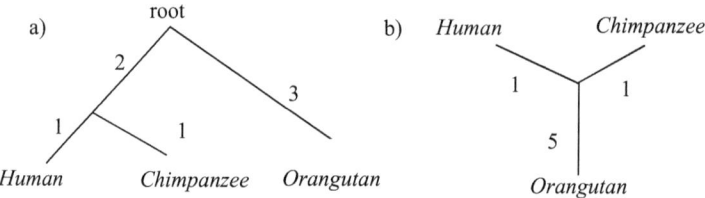

Figure 2. Phylogenetic trees of *Human,,Chimpanzee and orangutan*, a) a rooted tree, b) an un-rooted tree

There are many different algorithms to infer phylogeny from a given dataset. Based on different features of the sequence data, there are two main groups of algorithms, namely distance based and character based methods. Distance based algorithms require distance information of each pair of sequences in the dataset and cluster the sequences together iteratively. A common example of this type of algorithm is neighbor-joining [31]. The other group of algorithms uses individual characters of a sequence (nucleotides or amino acids), and is more informative than neighbor-joining. Maximum parsimony searches for the tree containing the minimum number of mutations, while maximum likelihood and Bayesian methods use probabilistic models. The following table gives an overview of different kinds of algorithms and their applications. Among these, mrbayes [32] has been suggested to be the best for inferring phylogeny [33, 34].

Programs for phylogeny estimation

Method	Software	Advantages	Disadvantages
Neighbor joining	Phylip,PAUP*, MEGA...	fast	Information lost Depends on MSA quality
Maximum parsimony	Phylip,PAUP*, NONA,MEGA...	Relative fast	Only search the minimum of the mutational pathway
Maximum likelihood	Phylip,PAUP*,PAML,Phyml,FastML,GASP,ANCESCON...	Fully likelihood of the phylogeny under a given model	Slow for large dataset
Bayesian	MrBayes, BAMBE	Strong connection to ML and faster for large dataset	Must specify prior distributions for parameters

Table 2. List of different methods and their popular software applications used for phylogeny estimation. The advantages and disadvantages of each type of algorithm are briefly mentioned [33].

1.2.5 Reconstruction of the ancestral state of a protein

Phylogenetic trees of biological sequences not only provide insight into their evolutionary history, but also into their common ancestor. Ancestral state reconstruction was first proposed by Pauling and Zucherkandl in 1963 [35], however, artificial reconstruction of a DNA or a protein sequence was not possible at that time. The rapid development of biotechnology and bioinformatics has made reconstruction of ancestral DNA or protein sequences possible and practical. Due to the large availability of sequence data, novel phylogenetic inference methods and powerful computers, ancestral states can be reconstructed fast and reliably. Even artificial synthesis in the laboratory has become relatively inexpensive, which allows investigation of the evolution of structure and function and the discovery of unknown functions that have been lost during evolution.

Ancestral reconstruction has recently drawn significantly more attention. It is widely used to study evolutionary pathways, adaptive evolution and functional divergence. In the 1990s, the last common ancestor of a digestive ribonuclease of swamp buffalo, river buffalo and ox was successfully resurrected and a functional test showed that it degraded RNA at least as effectively as the extant proteins [36]. Subsequently, many other ancestral proteins of various protein families such as vertebrate rhodopsin [37], elongation factor EF-Tu [38], chymase proteases [41], Tc1 transposons [39], and steroid hormone receptor [40] were resurrected and their biochemical properties were determined in the laboratory. For example, the ancestor of chymase has narrow substrate specificity as in alpha chymase [41]. The ancestor of eubacterial EF-Tu was a thermophile, not a mesophile or hyperthermophile, because its temperature optimum was 55-65°C [38]. The ancestor of rhodopsin in birds and other dinosaurs supported dim-light vision, which suggests that the first dinosaurs might have been nocturnal rather than diurnal [37, 42, 43]

Figure 3 shows a flow chart of an ancestral protein resurrection strategy. The first two steps are computational reconstruction work, and the output is normally a predicted protein sequence. The last 3 steps are required for the laboratory to produce an ancestral protein. The quality of a reconstructed phylogenetic tree is vital for the accuracy of the predicted ancestral protein. There are many factors that can affect the phylogeny, such as the sequences used, alignment algorithms, phylogeny reconstruction methods, and the evolutionary model used. More detailed information about how to infer a robust phylogeny is described in the Material and Methods section.

Maximum parsimony, maximum likelihood and Bayesian inference are the most widely used methods for ancestral protein reconstruction. Maximum parsimony [44] was first used in phylogeny-based methods for ancestral reconstruction. For closely related sequences it is effective and generally accurate. For example, in Hillis' study [43], 98.6% of all ancestral states were correctly reconstructed by parsimony. However, parsimony is a very crude evolutionary model, which assumes all evolutionary changes occur at equal rates. Even with weighted parsimony, which takes into account the observed variation of evolutionary rates, this method can still be problematic. This is because the maximum parsimony tree, which is completely determined by the reconstruction with a minimum number of mutations, does not fit real evolutionary processes [43].

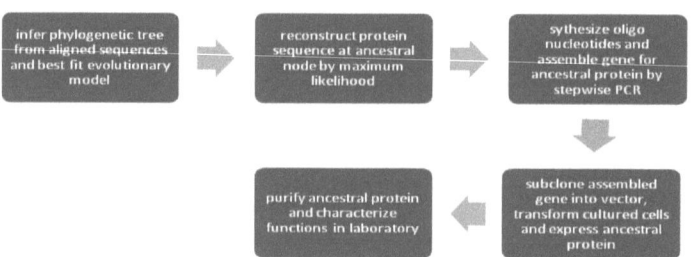

Figure 4. Flow chart of stages required for ancestral protein resurrection. Protein sequences are favored over DNA sequences because they are less noisy. Maximum likelihood or Bayesian inference methods are used to build the best phylogeny [43]

In contrast to parsimony, maximum likelihood [45] and Bayesian inference methods [46] result in a more reliable and accurate ancestral reconstruction [43]. These methods are based on a more realistic evolutionary model, in which multiple mutation events at the same site are taken into account, and all possible evolutionary pathways that are compatible with the data are considered. Compared to parsimony estimation, maximum likelihood and Bayesian methods are more accurate, especially for a highly-divergent set of sequences. Ancestral states that are ambiguous under parsimony can be estimated well by maximum likelihood.

1.2.6 Investigation of the evolutionary pathway

Evolutionary information is encoded in genetic material. Mutations accumulate from generation to generation and may cause loss or gain of functions. By using comparative proteomics and phylogenetic reconstruction, tracing evolutionary history has become possible and is now an important part of protein studies. On one hand, sequence elements important for function or structure in present-day proteins can be detected, which is important for studying selection and adaptation. On the other hand, information about the creation, expansion and extinction of both proteins and species can elucidate the evolutionary mechanisms responsible for the incredible diversity of life on earth.

1.2.7 Structure comparison

The PDB database provides a large number of high resolution three dimensional crystal structures of various proteins. Searching for kinesin structure, for example, gives 77 different kinesin structures in different states, such as nucleotide-free state, ADP binding, AMPNP binding, and even in complex with a microtubule. This enables a direct study of the structures by comparing the same protein in different states. Many important features, such as the structural conformation and protein-protein interactions can be discovered in this way. However, the comparison may be difficult because of the high complexity of the protein structure.

With the help of bioinformatics, similar structures can be aligned and visualized. The structural alignments can expose the positions, where potential conformational changes take place. These regions are normally under purifying selection. Thus, a local high sequence similarity can be observed. This makes it possible to use comparative analysis to study functional regions of proteins and to make predictions about conformational changes.

1.3 kinesin-1 project

1.3.1 kinesin super-family

Since kinesin was first purified and named in 1985, over 3,500 kinesin research papers have been published. Various types of kinesins have been detected and the kinesin super-family is thought to be subdivided into 14 sub-families, which was shown by the work of Lawrence et al. [14]. This result was supported by the work of Miki et al. [16] who analyzed over 600 kinesin sequences. However, depending on the number of kinesins sampled and the methods used for tree construction, the phylogenetic trees can differ substantially. For example, the sister group most closely related to the kinesin-1 family is kinesin-3 (formerly KRP85/85) in one tree (Figure 5) and kinesin-14 (formerly C-terminal kinesins) in another [14]. The aberrant kinesin SMY1 of *Saccharomyces cerevisiae* is a distant outgroup for all kinesins in one reconstruction [15] but a member of the kinesin-1 family in another [14].

As more and more genomes have been sequenced and have been made available in the public databases, the number of kinesin sequences has increased to over 2,000. In this project, the kinesin sequence database is expanded and a new phylogenetic tree of the entire kinesin super-family is constructed using an evolutionary model based maximum likelihood algorithm. This allows the previously defined kinesin families and their relationships to be tested with a much larger dataset and with more realistic phylogenetic models.

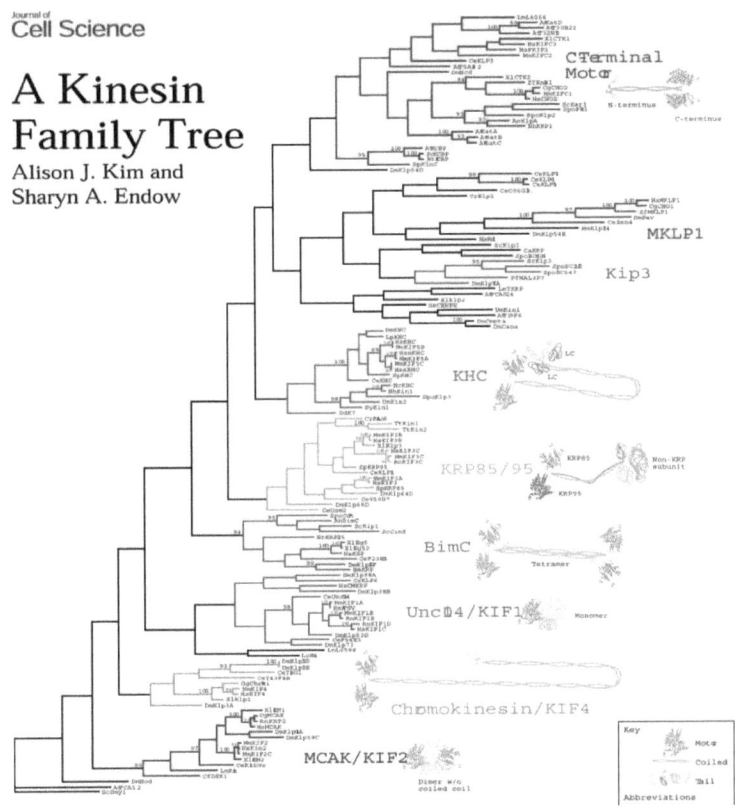

Figure 5. The tree shown above [15] was from a kinesin motor domain sequence alignment of 155 kinesin proteins from 11 species using the heuristic search method of PAUP* v4.0b10 [47], a maximum parsimony program, with random stepwise addition and tree bisection reconnection (TBR). The species included are a protist (*Plasmodium falciparum*), a yeast (*Saccharomyces cerevisiae*), two invertebrates (*Caenorhabditis elegans*, *Drosophila melanogaster*), a vertebrate (*Homo sapiens*) and a higher plant (*Arabidopsis thaliana*). The tree is one of two optimal trees that were found in 600 tree-building trials and has an overall length of 17,867. It is arbitrarily rooted using ScSmy1 as an outgroup to all kinesin proteins. The numbers adjacent to nodes are the percentages of 1,810 heuristic bootstrap replicates in which the indicated protein groups were found. The new names of the kinesin groups are shown with the former names in parentheses.

Until now, the classification of kinesin sequences has been based on phylogeny. For new kinesin sequences, using phylogeny to assign them to their corresponding groups is time-consuming and is strongly dependent on the quality of the phylogenetic tree. However, the phylogeny can vary according to the size of the dataset and range of taxa sampled.

A phylogenetic tree with a large number of kinesin sequences enables the creation of reliable character state models for each kinesin sub-family in order to classify new kinesin sequences quickly and correctly. An automatic classification tool is implemented in this project and integrated in the web server to make it accessible to kinesin researchers.

Together with the standard evolutionary tree of life, the kinesin distribution in organisms the evolutionary history of the kinesin is investigated briefly. When did a new kinesin group come into existance? How many common kinesin groups are there in one particular taxonomic group? How many kinesin groups differ among taxa. All of these questions are addressed in this project.

It is known that the motor domain of kinesin is highly conserved. A previous study has identified eight functional and structural motifs. However, because this analysis was based on only 106 sequences, it is questionable whether these motifs will be detectable as the number of sequences increases. In this project, a relatively robust set of motifs within the kinesin motor domain has been determined. It is used to predict functional and structural elements of the kinesin motor domain.

1.3.2 kinesin-1 sub-family

Kinesin-1, formerly called conventional kinesin, comprises kinesins involved in the transport of cargo through the cytoplasm. It is known that kinesin-1 can be found in all cell types and is expressed throughout cell development. Most kinesin-1s are located in the cytoplasm without binding cargo, while some transport various cargoes toward microtubule plus ends. Experimental tests with kinesin-1 antibodies have shown the inhibition of movement of tubular lysosomes, Golgi-derived transport vesicles, membrane bounded pigments and intermediate filament networks. Inhibition of kineisn-1 mRNA with complementary antisense oligonucleotides inhibits transportation of various proteins in axons [48-51].

In the presence of ATP, kinesin-1 can bind to microtubules for movement; however, the mechanism of converting energy from ATP hydrolysis into mechanical force is still unknown.

Kinesin-1 is a dimer formed by two identical chains, each chain consisting of a heavy chain, a coiled coil stalk and a light chain. The heavy chain is composed of the motor domain, normally about 325 amino acids long, and a short neck linker, 10 to 15 amino acids long, which is directly connected with the motor domain and binds the coiled coil stalk. The coiled coil stalk is linked to the tail region formed by the light chain, which is involved in cargo binding and appears to have regulatory functions.

1.3.3 Dichotomy of kinesin-1 in phylogeny and in motility

The development of in vitro motility assays, combined with very sensitive displacement and force measuring apparatus, has enabled direct monitoring of kinesin-1 motility in cell free assays by observation under the light microscope.

Motile properties of kinesin-1 from different species have been determined. Intriguingly, it was found that all tested animal kinesin-1 are slow motors ($\sim 0.6 \mu m\ sec^{-1}$) with comparatively low ATPase rates in the motor domain (k_{cat} 60-80 sec^{-1}) [52], whereas the fungal kinesins are "fast" motors ($\sim 2.5\ \mu m\ sec^{-1}$) with high ATPase activity ($k_{cat} \sim 260\ sec^{-1}$) [53].

This motor dichotomy clearly matches the dichotomy seen in the kinesin-1 phylogenetic tree, where kinesin-1 sequences from animal species and from fungal species are separated into two clades (Figure 6). This raises a number of fundamental questions: Why do fungi need fast kinesins and animals slow kinesins? Which domains or features of the motor determine moving speed? Can specific sites or motifs be identified that are associated with the different kinesins or determine the kinetic properties of the kinesin motor? How does a faster kinesin evolve into a slower one? Is it possible to convert a slow motor into a fast motor by making targeted changes in the primary sequence?

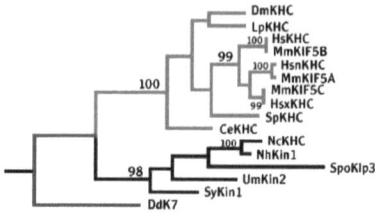

Figure 6. The kinesin-1 family tree built by Kim and Endow [15] using 14 sequences is clearly subdivided into an upper 'animal' branch (green) and a lower 'fungal' branch (black).

1.3.4 Previous laboratory research on motility

In the past decade, the enzymatic and kinetic properties of kineisn-1 of both fungal species and animal species have been characterized; some of them are listed in (Table 3). The tree groups ascomyceta, basidiomyceta, zygomyceta together, although they are actually rather evolutionary divergent. This is also seen among animals from highly divergent groups including insecta, vetebrata and mollusca. All five fungal kinesins show gliding velocities of between 1.8 and 3.4µm sec^{-1}, while all animal kinesin-1s show gliding velocities in the range 0.4-0.9µm sec^{-1} (Table 3).

Fungal kinesin-1			
Species	Group	Speed	Reference
Neurospora crassa	Ascomyceta	2.5µm sec^{-1}	[53]
Aspergillus nidulans	Ascomyceta	2.01µm sec^{-1}	[54]
Syncephalastrum racemosum	Zygomyceta	2.1–3.4µm sec^{-1}	[55]
Ustilago maydis	Basidiomyceta	1.8µm sec^{-1}	[56]
Animal kinesin-1			
Drosophila melanogaster	Insecta	0.9µm sec^{-1}	[57]
Homo sapiens	vetebrata	0.45µm sec^{-1}	[58]
L. pealii	Mollusc	0.5µm sec^{-1}	[59]

Table 3. List of kinesins that have been experimentally studied their taxonomic group and velocity.

To address the question of how structure determines motor velocity, there have been many experimental attempts over the years. For example, chimeras of *Neurospora* kinesin-1 and human kinesin-1 were generated where the neck and hinge domains of *Neurospora* kinesin, which are important for motor function (kallipolitou et al. 2001) [60], were replaced by the corresponding human kinesin domains (Figure 7).

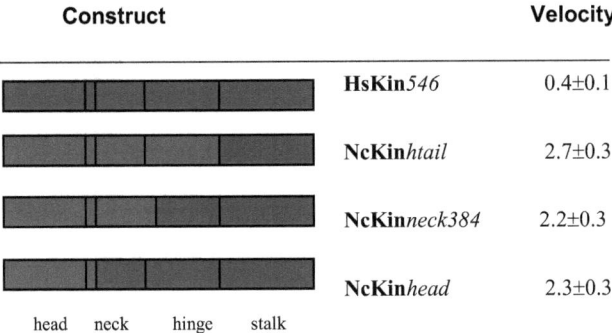

Figure 7. Motor speed of chimeric constructs as measured in gliding assays. Red indicates Neurospora kinesin-1 and green to human kinesin-1. The neck construct was extended by 5 amino acids into the hinge since the hinge tryptophan 384 has been demonstrated to be crucial for dimerization of the Neuropora kinesin-1 neck [58].

These experiments indicate that the gliding velocity is exclusively determined by the motor domain (unpublished observations by F.Bathe, Ph.D thesis, 2004), but not the neck and hinge domain, which were demonstrated to be important for motor activity by many studies [55, 58, 60]. Replacement of the neck, hinge and stalk domains of slow motors in the human kinesin-1 (see Figure 8) did not affect kinesin velocity.

Another attempt to gain preliminary insights into structural determinants of motor function invloved generating chimeras of human and *Neurospora* kinesin-1 by subdividing the head into a front region and a back region (Figure 8), which was suggested by a three dimensional crystal structure comparison of kinesin-1.

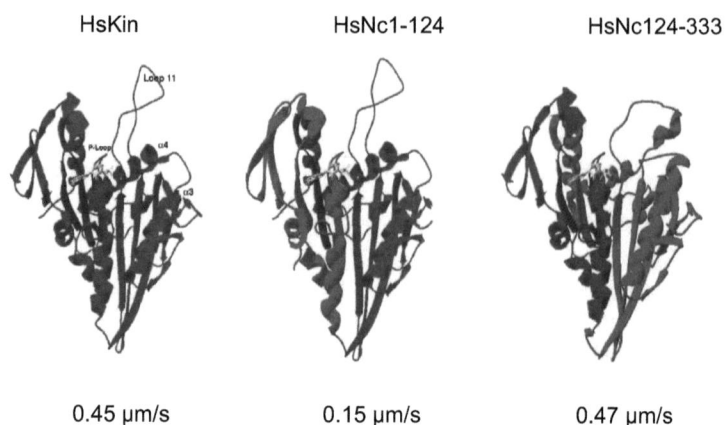

Figure 8. Motor domain chimeras of human kinesin-1 (green) and *Neurospora* kinesin-1 (red). The rest of the construct was derived from human kinesin

This experiment showed that a simple rearrangement of structural domains between a fast and a slow motor (third chimera in Figure 8), does not result in higher motor velocity. In contrast, the second chimera in Figure 8 actually slows down the motor (unpublished; A. Kallipolitou, PhD thesis, 2001).

Since the motor domain is essential for kinesin motility, comparison of motor domain sequences could provide insights into how heads regulate speed. Amino acid positions that differ between animal and fungal groups, but are highly conserved in either group have been identified and are thought to be important for motor functions. In an alignment of 11 animal and 8 fungal kinesin-1 motor domain sequences 40 sites were identified by an absolute conservation (100%) cut-off. Among these sites, one serine to glycine exchange in fungal kinesin-1 is particularly conspicuous (Steinberg and Schliwa, 1995), which gives the switch-II region of the sequence the character of a p-loop (highlighted in yellow in Figure 9).

Animals: ... VDLAGSEKVSKTGA...
Fungi: ... VDLAGSEKVGKTGA...

Figure 9. Alignment pattern of the switch-II region.

A point mutation was generated at this site. By introducing a SKT motif into *Neurospora* kinesin, its gliding velocity was reduced to 73% of the wild-type velocity; however, the reverse experiment, in which the GKT motif was introduced into *Drosophila* kinesin-1, did not change the motile property of motor (unpublished; U. Majdic, PhD thesis, 1999). This attempt showed that a single point mutation is unable to convert the slow *Drosophila* kinesin into a fast motor.

All of these experimental attempts indicate that a more sophisticated method is needed to address determinants of kinesin motor speed.

1.3.5 Attempts by Bioinformatics

1.3.5.1 Comparative approaches

The application of bioinformatic approaches is needed as increasingly more kinesin sequence data become available. These approaches serve to facilitate analysis of kinesins at the sequence level and provide insights into the relationship between the motility and the structure of kinesins.

Comparative approaches have shown that conserved regions of the primary sequence are important for function [61]. There are two types of conserved residues that are particularly important for kinesins. The first group of conserved residues is important for protein function, forming the ATP binding sites and the microtubule binding sites. The second group of residues is related to protein structure, such as folding into a 3D structure.

While common functions of kinesins can be revealed by comparison of the entire kinesin super-family, comparisons within the kinesin sub-families can reveal group specific functions, which can be inferred from conserved group-specific residues. Assuming that determinants of different kinesin velocities are at least some of the group specific residues, it is of interest to determine residues that are highly conserved in the fungal kinesin group and differ from corresponding conserved positions in the animal kinesin group (Figure 10).

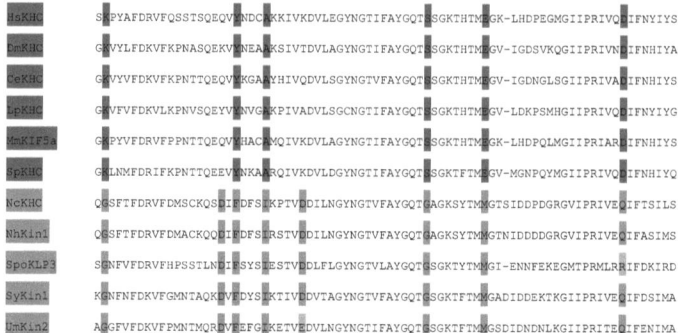

Figure 10. Example of a sequence comparison of an 80 amino acid segment from 6 animal (red) and 5 fungal (green) kinesins. Conserved positions are highlighted in red and green and exceptions are shown in light blue. Eight sites were highly conserved in all fungal kinesin-1. Six of them are highly conserved in all animal sequences but with different residues than the fungi. Two others are less conserved

Sequence comparison of 11 animal and 8 fungal kinesin-1 protein sequences reveals 40 variable sites in the head domain region. Site-directed mutagenesis of one of these sites even reduced the velocity of wild-type *Neurospora* kinesin, which demonstrates its functional significance. However, there are several shortcomings to the above analysis. The sample size of sequences from both the animal and the fungal group should be large enough to avoid incorrect phylogenetic inferences. The identified residues are potentially important for function or structure, but there still can be residues that are less conserved in one group but nevertheless differ from the residues found in other group and which interact with the highly conserved group-specific residues. Including the three dimensional atomic structural information is imperative in this case to obtain a complete set of potential determinants. Furthermore, these residues can be divided into several groups depending on their spatial interaction and used in turn to guide laboratory generation of mutants to see for example whether human kinesin-1 can be converted into a faster motor by introducing of a group of fungal-specific residues.

1.3.5.2 Resurrecting ancestral kinesin-1 proteins

Comparative approaches can reveal functionally important residues, including residues that control kinesin velocity. However, testing and confirming these predictions will be clearly both labor and cost intensive.

With the increasing number of genetic sequences and sophisticated evolutionary model-based phylogeny inference algorithms, reconstruction of reliable phylogenies of protein super-families is becoming increasingly common and crucial for the study of molecular evolution. It not only exposes the relationship of extant proteins, but also makes it possible to reconstruct ancestral proteins and to apply statistical methods to estimate past evolutionary changes to a sequence that occurred at any internal node in the phylogenetic tree.

Resurrection of ancestral kinesin-1 is considered more promising, less time consuming and less cost intensive than sequence optimization. On one hand, ancestral kinesin-1 can be resurrected in the laboratory and used for direct testing of kinesin catalytic properties such as ATPase activity and motility; on the other hand, ancestral kinesin sequences enable the mapping of any site on the phylogenetic tree and show the timing and directionality of sequence changes (Figure 11).

The estimation of a robust kinesin-1 phylogeny is the most important step in the resurrection of an ancestral protein. It is crucial that as many kinesin-1 sequences as possible are included in the tree construction. The more sequences that are used the more reliable the resulting phylogenetic tree will be. For example, if kinesin-1 exists in primitive organisms such as coelenterates (Hydra and Nematostella), the sponge Reniera, the placozoan trichplax or any representative of the choanoflagellates, which is believed to be the base of animals, its sequence can help to extend the phylogeny and increase the accuracy of ancestral kinesin reconstruction.

Ancestral reconstruction programs such as GASP [62], gapped ancestral sequence prediction, infer the ancestral sequence at each internal node in the phylogenetic tree. These intermediate sequences are extremely important in exposing the evolutionary history of kinesin. The most interesting ancestral sequences are those representing the common ancestor of all extant animal species, the common ancestor of all extant fungal species, and the common ancestor of both animals and fungi. Resurrecting the tree of ancestral kinesins and testing their biochemical properties can help us understand whether the ancestor of fungi also had a fast kinesin and if the ancestor of animals had a slow one. The answers to these questions are important for the identification of specific sequence changes that convert a fast kinesin into a slow one.

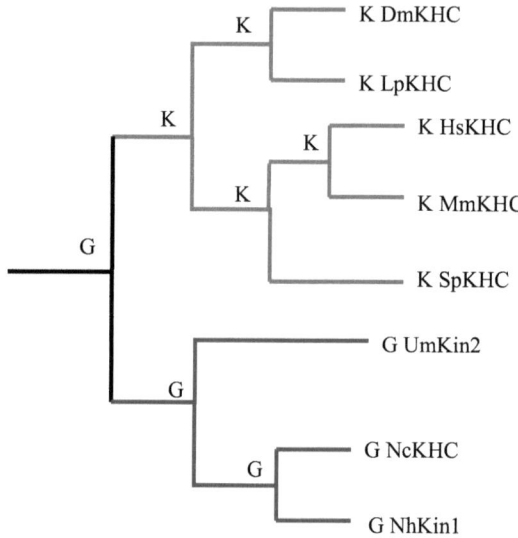

Figure 11. Simple example of site evolution mapped on a phylogenetic tree. At this site, a derived lysine (K) residue is present in animal kinesin-1 sequences (green branch), while the ancestral glycine is present in fungal kinesin-1 sequences (blue branch). Amino acid states of the root and internal nodes were inferred by the ancestral reconstruction algorithm.

1.3.6 Goals of the work

The number of available kinesins has increased by over 2500 in the public database. One intention of the present work is to comprehensively analyze the kinesin super-family using bioinformatic approaches based on a large dataset. To this end, a method for automatic detection and classification of kinesins was developed; amino acids and motifs that crucial for kinesin's functions and structures were predicted and the evolutionary history of kinesins was investigated according to the distribution of kinesin sub-families in various species.

Another aspect of the work is to infer a big and high-quality phylogenetic tree of the kinesin super-family and recheck the standard kinesin sub-family classification scheme. Maximum likelihood inferring method was used instead of neighbor-joining method to avoid information loss. A new method was developed to test the reliability of inferred tree, because the classic bootstrap test was very CPU-time consuming for large datasets. Based on the new phylogenetic tree, the classification scheme was confirmed.

Furthermore, this work aims to explain why fungal and animal kinesin-1s have very different velocities. Programs were implemented for detecting group-specific residues. The mapping of these residues onto the fungal and animal three-dimensional crystal structures (1BG2 and 1GOJ) has led to the discovery of several structural changes from a closed to an open conformation of the motor domain. Possible combinations of residues that could impact the velocity were predicted.

In addition, ancestors of fungal and animal kinesins were reconstructed in order to understand the velocity difference by studying the evolution of velocity

At last, a kinesin web-server was constructed. It automatically detects, classifies new kinesins and stores in the database. It provides not only useful tools for analyzing kinesins from sequence to structure, but also tools that can be applied to any protein datasets.

26

2 Materials and methods

2.1 Data collection

2.1.1 NCBI RefSeq database

The RefSeq database, started on October 9th, 2002, is a continuing project of the National Center for Biotechnology Information (NCBI), which aims to provide a non-redundant collection of well-annotated DNA, RNA, and protein sequences from diverse taxa.

Compared to other databases, RefSeq provides unique, curated sequences in addition to rich and accurate information. Sequences from the RefSeq database are derived from GenBank records, however, unlike GenBank, which is an archive of sequences and annotations supplied by original authors and cannot be altered by others, each RefSeq represents a synthesis of the primary information that is generated and submitted by a person or group. This results in an accurate annotation of each molecule with the organism name, strain, gene symbol and protein name by either NCBI staff or extensive collaboration with authoritative groups.

RefSeq sequences can be easily accessed by many web sources, such as BLAST results, Conserved Domain Database (CDD) [63], HomoloGene, UniGene and Clusters of Orthologous Groups of protein (COG) [64], which implies that a huge information network can be linked via RefSeq records.

RefSeq is not only accessible online, but also has a flat file format, which can be downloaded via FTP to a local workstation. The records of RefSeq have a similar format to the GenBank records from which they are derived, but with many novel attributes, such as a unique accession prefix, a comment field that indicates the RefSeq status and a source of the sequence information.

RefSeq updates are provided daily, including new entries added to the collection or records updated to reflect sequence or annotation changes. The daily update file is also provided at the NCBI FTP site.

RefSeq is a unique, accurate, information-rich, easily accessible and up-to-date database. Its records can be found widely in many other useful web sources other than NCBI itself, which makes research or analysis of its biological sequences easy and reliable [65].

The Januray, 2009 RefSeq collection, release 33, includes sequences from 7,773 distinct taxonomic identifiers, which range from viruses to bacteria to eukaryotes. It represents chromosomes, organelles, plasmids, viruses, transcripts, 6,413,124 proteins, 2,226,548 genomic sequences, and 1,685,610 RNAs (Table 4). Every sequence has a stable accession number, a version number, and an integer identifier (gi identifier) assigned to it.

Number of taxa: 7,773		
	Number of Accessions	
Genomic	RNA	Protein
2,226,548	1,685,610	6,413,124

Table 4. Summary of data in RefSeq release 33.

The sequence data used in this project were derived from the protein database of RefSeq release 33[66].

2.1.2 Kinesins in the RefSeq database

The volume of biological sequences has exploded thanks to novel sequencing technologies, like nanopore technologies and pyrosequencing, which increase sequencing speed at least 100-fold over the traditional Sanger method.

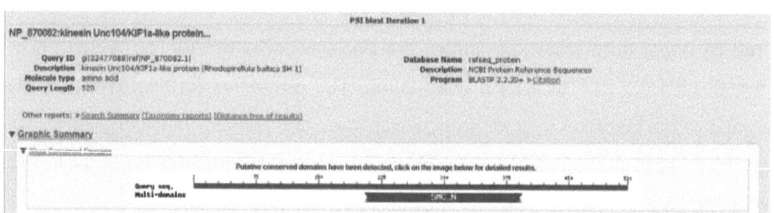

Figure 12. Sequence example with incorrect kinesin annotation. gi|32477088|ref|NP_870082.1| is annotated as kinesin Unc104/KIF1a-like protein. However, the conserved domain search shows that it

contains a SMC_N domain, located in the middle of the sequence, but no kinesin motor domain. This indicates that protein sequence gi|32477088|ref|NP_870082.1| cannot be a kinesin.

A search for "kinesin" in the NCBI protein database gives 11,822 hits, including 4,501 hits in the RefSeq database, many of which are duplicates or incomplete. Some of them are even incorrectly annotated (Figure 12). For this reason, a more sensitive search was applied to obtain a high-quality kinesin dataset.

2.1.3 PSI-BLAST search against the RefSeq database

A PSI-BLAST [67] search for kinesin homologues against the RefSeq database was applied starting with seven selected query kinesins: two from the metazoan group, *Monosiga brevicollis* (Mb) and *Homo sapiens* (Hs), one from the Amoebozoa group, *Dictyostelium discoideum*(Dd), one from the plant group, *Arabidopsis thaliana* (At), and three from the fungal group, *Neurospora crassa* (Nc), *Ustilago maydis* (Um), and *Yarrowia lipolytica* (Yl). Using sequences from various groups and making several runs of PSI-BLAST can help to find a more diverse range of kinesin homologs.

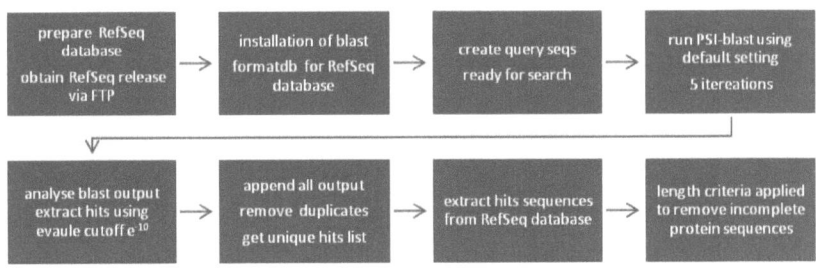

Figure 13. Flow chart of kinesin search against the RefSeq database.

The detailed search procedure is shown in Figure 13. Five iterations were run for each PSI-BLAST search and a maximum of 10,000 best hits were taken from the 5th iteration. An e-value cutoff of e^{-10} was used as the criterion for selecting significant hits. Finally, duplicates were deleted and incomplete sequences were removed using a minimum length requirement of 300 amino acids. That is, sequences shorter than 300 amino acids were considered incomplete and excluded from further analysis.

2.2 Classification of kinesin sequences

Kinesins are a large protein family and studies have shown that there are about 14 different kinesin sub-families, which vary in both structure and function. However, a fixed nomenclature is not used when annotating kinesins in the public database, which makes it difficult to group kinesins into sub-families. Classification of kinesins has been a stumbling-block for many analyses of particular kinesin sub-families.

Many classification criteria have been developed in the past years, such as sequence similarity classification, functional classification, structural classification, profile classification, phylogenetic classification, etc. Kinesins are multi-domain proteins that typically share high similarity in the motor domain region, but otherwise vary in shape and function. Classification based on simple criteria such as similarity or structure can results in a loss of accuracy because of shared common features between groups. In this case, the profile classification and phylogenetic classification methods can better classify the kinesins.

2.2.1 NCBI CDD classification method

NCBI provides a conserved domain database (CDD), which is a collection of multiple sequence alignments for ancient domains and full-length proteins in the form of position specific score matrices (PSSMs) [68]. The reverse position-specific BLAST (RPS-BLAST) [63] search method is used to scan the CDD database for query sequences and e-values are obtained similarly to PSI-BLAST. Unlike other models (like pfam, smart or COG), CDD detects evidence for duplication and functional divergence in domain families by using phylogenetic information. Kinesin has a structure as described by a set of 14 explicit PSSM models. When scanning a kinesin protein query sequence against CDD kinesin models, a region of query sequence may hit more than one overlapping motif. The hits with the best score or lowest e-value provided by one of the 14 kinesin models suggests which sub-family the query sequence most likely belongs to. When the best hit does not match any of the 14 models, the query sequence is considered as a kinesin homolog but not a true kinesin because it lacks crucial amino acids required to be a functional kinesin.

2.2.2 Hidden Markov model classification

Since there is no sub-family information for kinesin sequences available, phylogeny can provide hints about which sequences have the same evolutionary history and can be grouped together. Theoretically, the phylogeny of a protein family can show sub-family information of every sequence by placing them into the same clades. Practically, it's hard to build an error-free phylogeny due to incomplete taxon sampling; however, by focusing on the main clades with good support values, one can extract related sub-families of a protein super-family. Kinesin sequences were classified by CDD classification at first, and then a maximum likelihood phylogeny tree was built. Clades containing only sequences with the same class identification were selected, and an alignment for each group was built, which was then used for creating a hidden Markov model with the hmmer application [69,70].

Similarly to CDD, hmmer provides a hmmpfam search tool, with which a query sequence can be scanned by hmm models and an e-value obtained for each hit. The best hit is considered as the sub-family that the query sequence belongs to.

Hmmer classification was used as an extra criterion to increase the accuracy of the sub-family classification.

2.2.3 Significance test

In cases where multiple hits have similar e-values, a significance test is performed. Assuming that the best hit has e-value e1, and the second best hit has e-value e2, two test methods were used, the e-value thresholds cutoff methods and the likelihood ratio test (described below).

2.2.3.1 E-value threshold cutoff

The e-value threshold cutoff is an easy way to determine a significant hit. A hit is considered significant when the difference between two e-values is above a pre-selected threshold. For example:

$$diff = e1 - e2$$

$$f(diff) = \begin{cases} \text{not significant,} & diff < threshold \\ \text{significant,} & diff \geq threshold \end{cases}$$

2.2.3.2 Likelihood ratio test

The likelihood ratio test (LRT) is a statistical test of the goodness-of-fit between two models. That is, this test compares a relatively more complex model to a simpler model to test if the complex model better fits a particular dataset. The LRT is only valid when two hierarchical nested models are compared. Nested means that the complex model must differ from the simple model by addition of one or more parameters [71]. The LRT begins with a comparison of the likelihood scores of the two models as follows:

$$LR = 2*(\ln L1 - \ln L2)$$

This LRT statistic approximately follows a chi-square distribution. To determine the critical value of the test statistic from standard statistical tables, one must first determine the degrees of freedom, which is the same as the number of additional parameters in the complex model. By using the critical value, one can interpret whether the difference in likelihood scores between two models is statistically significant [72].

We take that the best hit as the null hypothesis and the second-best as the alternative hypothesis,

	lnL
H0	E1
H1	E2

The CDD model differs from the hmmer model by two additional parameters μ (location) and λ (scale) [70]. Thus, the degrees of freedom are two. For each hypothesis there is one e-value to be determined. Therefore,

$$LR = 2*(E2-E1), E2 > E1$$

By comparing LR and the critical value C (standard statistical value) in the table of the chi-square distribution with two degrees of freedom, one can use the following criteria to accept or reject the null hypothesis. Normally, a 5% significance level (P=0.05) is used.

If C(P=0.05) <= LR, accept H0

If C(P=0.05) > LR, reject H0

The chi-square distribution is shown in Figure 14 with the blue highlighted area indicating the 5% significance level.

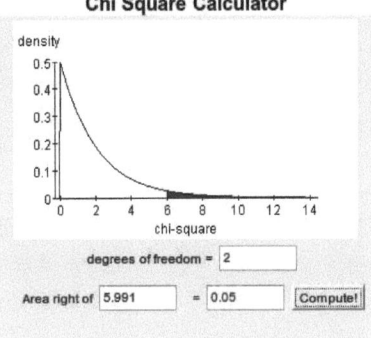

Figure 14 Chi-square distribution for P=0.05 and degree of freedom=2. The highlighted region indicates that the critical value is greater than 5.991and is in the 5% significance range.
(http://www.stat.tamu.edu/~west/applets/chisqdemo.html)

2.2.4 Automatic classification of kinesin sequences

Figure 15 shows the flow chart of the automatic kinesin classification method. A query sequence is classified with CDD classification and hmmer classification separately. A significance test is done afterwards. The more significant result is assigned to the query sequence. The annotation of kinesin sub-family name is described blow in section 2.2.5.

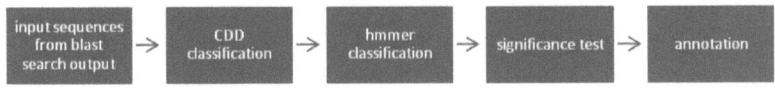

Figure 15. Flow chart of the automatic kinesin classification application. Query sequences are classified using both NCBI CDD criteria and hmmer criteria. Results are tested with a significance test and the sequence is then assigned to a kinesin sub-family.

2.2.5 Nomenclature of kinesins

Kinesin sequences were renamed with their class name based on the standardized kinesin sub-family names published in 2004. Since the CDD database did not use the standard nomenclature, a match of CDD database names to standard class names were applied for each kinesin sequence after an automatic classification approach (Table 5). Hmmer profiles were named after the standard class names.

CDD name	Group	Member example
kisc_khc_kif5	1	*N. crassa* KHC
kisc_kif3	2	*H. sapiens* KIF3
kisc_kif1a_kif1b	3	*C. elegans* Unc-104
kisc_kif4	4	*H. sapines* KIF4
kisc_bimc_eg5	5	*H. sapiens* KSP (HsEg5
kisc_kif23_like	6	*H.sapiens* MKLP1
kisc_cenp_e	7	*H. sapien* CENPE
kisc_kip3_like	8	*S. cerevisiae* KIP3
kisc_kif9_like	9	*M. musculus* KIF9
kisc_kid_like	10	*M. musculus* KIF22
kisc_klp2_like	11	*M. musculus* KIF15
kisc	12	orphans
kisc_kif2_like	13	*M. musculus* KIF2
kisc_c_terminal	14	*D. melanogaster* Ncd

Table 5. Match table of CDD names of kinesin sub-familes and standard kinesin nomenclature.

2.3 Automatic update of the kinesin database

Hundreds of new sequences are made available online daily, which makes it necessary to search the database regularly in order to maintain an up-to-date dataset. This is also the case with kinesins.

The RefSeq database provides a daily update file, which makes it easy to update kinesin data.

The daily update file is available at: [ftp://ftp.ncbi.nih.gov/RefSeq/daily/]. This file is used to create a daily database for a blast search with the formatdb [26]. A psi-blast search similar to the one described above is used to obtain new kinesin hits. After the refinement, a fasta format sequence file is generated and is ready for classification. All kinesin homologues are removed and only the true kinesins are assigned with a unique class name and saved in the database.

2.4 Multiple sequence alignments

For the kinesin sequences dataset, the clustalw program [73], version 2, was used to generate alignments. Muscle version 3.6 is an alternative choice, which generally performs better than other alignment programs [74]; however, clustalw provides better alignment quality when comparing several motifs within kinesin sequences.

Alignment of all kinesins was generated by the clustalw2 using the default settings. An automatic refinement of the alignment was done with Rascal [75] with five iterations.

Alignments were generated for each kinesin sub-family to perform a sub-family orientated analysis, such as a specific motif pattern search, and a sub-family phylogenetic ancestral protein sequence reconstruction. Alignments for each pair of sub-family combinations were made by profile alignments to examine variation between sub-families, such as sub-family specific amino acids, motifs and structures. Alignment of the entire kinesin super-family was used to create a super-family phylogeny to reveal the relationship between sub-families and the evolutionary history of kinesins.

The alignment of all kinesins is updated when new kinesins become available by a sequences-profile alignment. The alignments are accessible at the website [www.bio.uni-muenchen.de/~liu/kinesin_new/]

2.5 Comparative proteomics

Amino acids are the building blocks of proteins. The proteins aid in controlling almost every biochemical reaction within the cell. Understanding functions of proteins and how they work is the key to addressing questions about how cells function and evolve.

With computational approaches and the explosion of biological data resources, prediction of protein functions and structures is no longer mission impossible. Comparative genomics is a new approach to understand how life changes over time on the molecular level. That is, to infer the evolutionary pathways of proteins by comparing their sequence and structural similarities and differences in different organisms and exposing the relationship between protein structure and function across various species.

2.5.1 Conserved amino acids in kinesin sequences

Proteins that have similar functions and structures in different species are expected share high sequence similarity. Thus, the conservation of sequence blocks across species can be used to identify regions of a protein that are important for function and structure. Kinesins share high sequence similarity in their motor domain and a blast search can successfully extract thousands of kinesin candidates from the millions of known proteins. Some regions of the protein sequence differ among kinesins that differ in function and these regions can be used for classification of kinesins into different sub-families.

Therefore, it is very important to identify the conserved positions in kinesins. These positions contain information about both functions and structures of kinesin such as the active center, protein-DNA interaction positions, protein-protein interaction positions, structure building etc.

The degree of conservation can be determined from protein sequence alignments. The following are guidelines on how to define and quantify conservation among species.

- How is conservation defined?

The calculation of conservation at the amino acid level uses an alignment of protein sequences. Assuming C is one column in an alignment of n proteins, conservation can be defined in various ways by using different features of amino acids. Three definitions of conservation are presented here.

- absolute conservation

    Conservation is defined as the occurrence of the same amino acid in the same position of proteins. Conservation of amino acid a is calculated as follows:

$$\text{Conservation}(a) = \frac{N(a)}{n}$$

Where $N(a) = \sum_{i=0}^{n} f(i)$ $f(i) = \begin{cases} 1, & C = a \\ 0, & C \neq a \end{cases}$

- Hydrophobic conservation

    Hydrophobicity and hydrophilicity are important amino acid features, which are important properties of protein structure and protein-protein interactions. At a given position in the alignment, the specific amino acids may differ among proteins, but still share the same hydrophobic properties. For example, hydrophilic amino acids, like serine and threonine, are used on the surface of soluble proteins, while membrane proteins normally have hydrophobic amino acids at the ends so that they can lock into the membrane. In kinesins, different hydrophobic amino acids used in building the active center of the ATP binding pocket can be treated as conserved. In this case, we need to modify the decision function f(a) to calculate the conservation of amino acids.

    $$f(a) = \begin{cases} 1, & \text{hydrophobicclass}(Ci) = \text{hydrophobiclass}(a) \\ 0, & \text{hydrophobicclass}(Ci) \neq \text{hydrophobicclass}(a) \end{cases}$$

    where the hydrophobicclass of an amino acid is based on its hydrophobicity and its occurrence on the surface and interior of the protein structure and defined as follows

    - (cvlimfw): hydrophobic => interiors
    - (rkedqn): hydrophilic => surface
    - (phygast): neutral   => neutral/both

- Polar conservation

    Polarity is a physical feature of amino acids. Polar conservation is calculated similarly as the hydrophobic conservation, but with a different function:

$$f(a) = \begin{cases} 1, & \text{polarclass}(Ci) = \text{polarclass}(a) \\ 0, & \text{polarclass}(Ci) \neq \text{polarclass}(a) \end{cases}$$

where the polar class of an amino acid is based on its polarity and defined as follows

- (gavlipfwm): non-polar
- (scnqyt): polar-uncharged
- (edhrk): polar-charged

- Which alignment should be used to calculate conservation?

After defining how conservation should be calculated, one must specify an alignment. The alignment can be built with a single sub-family, a combination of sub-families or a super-family. The choice of sequence breadth will vary depending on the questions investigated. For example, when studying ATP binding sites of the kinesin-1 sub-family, an alignment of the kinesin-1 sub-family should be used to calculate conservation. Afterwards, in order to compare conservation of ATP binding sites within kinesin-1 with the conservation of ATP binding sites all over the kinesins, a recalculation of the conservation with the alignment of all kinesin sequences should be performed.

With pre-created alignments for every combination of kinesin sub-families, the investigation of conservation of any important amino acids within various sets of kinesin sequences becomes an easy task.

2.5.2 Identifying conserved sequence motifs

Finding conservation of individual amino acids is important in revealing functional positions within a sequence. However, to better understand elements that are functionally or structurally important, we need to place these amino acids into larger sequence patterns, which are known as sequence motifs.

A simple and effective tool was developed to scan the alignment for potential motifs. This method is based on the conservation level of amino acids. A sequence motif in a protein sequence can either be a functional motif or a structural motif. A functional motif has a biological significance like DNA binding, protein interaction, activation of protein etc. A structural motif is formed by the three dimensional arrangement of amino acids and is important for determining protein structure. Functional or structural motifs are usually conserved within a particular protein set. They can sometimes vary in length and contain gaps. Each motif should contain at least one conserved amino acid. That is, finding a motif is the same as starting with a conserved amino acid and extending from both sides until a convergence criterion is reached. To avoid high computing time, a linear pattern scan algorithm with linear running time is employed (Kadane's algorithm) [107]. The motif scan program works as follows:

Setting up parameters:

1. A conservation cutoff, which defines the conservation level of a motif.

A lower boundary of the cutoff is defined as the average conservation of an alignment. Columns in the alignments with more than 90% gaps are excluded from the conservation calculation. A pattern, whose conservation level is above the cutoff, is considered a significant motif.

2. Gap length

A gap length is the maximum length of gaps allowed between two conserved amino acids or patterns, in order to define the endpoint of a motif. A position with conservation level less than the average is defined as a gap. In the motifs these gaps are represented as X.

Scanning the alignment:

1. The scan starts from the first position of a given alignment. The conservation level of each position in the alignment is calculated. The consensus amino acid is the amino acid that has the maximum conservation score for that column of the alignment.

2. Scanning the alignment

a) The conservation score and the maximally conserved amino acid are stored in an array in the 1st step.

b) The average conservation score is subtracted from every array element, so that the significant conserved positions have positive values, while the others have negative values. The motif finding problem is reduced into a maximum subarray problem in this way. The average conservation score can be user-defined. Otherwise, the average conservation score of the alignment is calculated automatically by the program. Empty columns (over 90% gaps) are excluded.

c) A motif starts from a positive position and extends by adding its right-hand neighbor. The motif is terminated when either the average conservation score of the current motif is less than the predefined conservation cutoff or the gap length exceeds the given maximum.

d) Non-conserved positions within a motif are indicated by small 'x's.

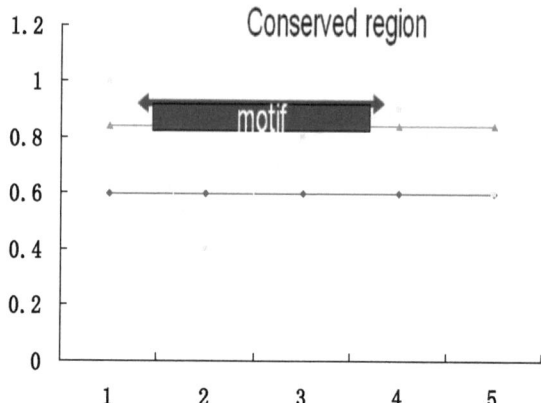

Figure 16. Example of the linear motif search in a protein alignment. The yellow line represents the maximum conservation at each site. The red line is the average conservation score of the alignment. The green line is the user defined conservation cutoff.

With this algorithm one can easily find motifs in a given sequence alignment using different conservation methods (absolute conservation, polar conservation or hydrophobicity conservation).

2.5.3 Finding known motifs in proteins

Motifs are important for functional and structural studies of proteins and many databases of common motifs have been established within the last decade. Many annotations of predicted, unknown proteins are based on motifs present in the sequence. Once a known motif has been found in a new protein, and this motif is shown to be conserved, then it is highly probable that this protein has a partial function or the same structure as others containing the same motif.

Unlike searching for motifs in a sequence alignment, a reverse motif search can find all sequences that contain a known motif pattern.

A known motif is given in the form of a regular expression notation [76]. It is then converted into a mysql [77] search pattern and finally used to scan the entire sequence database. The mysql implemented pattern-matching algorithm 'REGEXP' is used for the search [77]. The search returns all sequences that contain the given motif. The conservation score of each motif is then calculated using the sub-family alignment to which the sequence belongs.

2.5.4 Discriminating amino acids

It is known that different kinesin sub-families have various functions. The amino acid residues responsible for the divergence between sub-families are called discriminating amino acids. It is assumed that they help the protein perform sub-family specific functions rather than general biological functions common to all sub-families.

Discriminating amino acids can also be detected within a sub-family. For example, the fungal kinesin-1s move four times faster than animal kinesin-1s. A discriminating residue shows a significantly higher conservation level in one sub-group (for example fungi) than in other sub-groups (for example animals).

Assume that an alignment consists of two different sub-groups. While looking at one column of this alignment, a maximally conserved residue for each group can be determined. When the maximally conserved amino acids differ between groups and their conservation difference is significant, then this position is called a discriminating position and the amino acid residue used by one group is called a discriminating amino acid.

A computational automatic detection approach has been developed to find discriminating amino acids. It takes alignments of two or more sub-groups as input. The conservation score is calculated for each sub-group. When the conservation of any sub-group is over 80%, the residue is identified as a potential discriminating residue.

2.6 Phylogeny reconstruction

2.6.1 Methods

RaxML, a maximum likelihood implementation for phylogeny inferrence, was used to generate a large phylogenetic tree for the entire kinesin super-family. The tree was generated from a random starting tree, using the blosum62 matrix as the amino acid substitution model with a gamma-distributed model for rate heterogeneity. By default, four discrete rate categories were used [78].

2.6.2 Visualization

Phylogenetic trees created by mrbayes were visualized by using forester version 4.1 [79], software libraries for evolutionary biology and comparative genomics research, and FigTree version 1.2.1 [80]. Sub-families were highlighted with different colors.

2.7 Reconstruction of ancestral kinesins

Ancestral protein reconstruction is a method to infer the common ancestor of a set of sequences using a statistical model. Tracing the extant proteins back to a common ancestor is challenging because it is impossible to know the exact sequence changes that occurred on the evolutionary pathway from the ancestor to the current state. Currently, more realistic algorithms, such as the stochastic model, maximum likelihood, are being used to solve this problem. Given a phylogenetic tree, one can start from the leaves to calculate the likelihood of all possible changes of the internal nodes and then iteratively repeat this until root is reached. Therefore, the quality of the estimation should strongly depend on the given phylogenetic tree. GASP [62], gapped ancestral sequence prediction, version 1.3, and was used for kinesin ancestor reconstruction. GASP is advantageous because it can use a user-defined phylogeny in contrast to rather than other applications, which normally only use the alignment to estimate the phylogeny by their internally implemented algorithm. These algorithms create different phylogenies in general and consequently lead to inaccurate estimation of ancestors.

2.8 Structural comparison

Experimentally determined molecular structures of kinesin proteins were obtained from the PDB database. Each kinesin protein sequence was classified by the kinesin classification tool. Structures were visualized by RasMol [81]. The comparison was done for two kinesin-1 proteins: the *Neurospora crassa* motor domain 1GOJ and the *Homo sapiens* motor domain 1BG2.

Motifs, conserved amino acids and discriminating amino acids were mapped onto structures. An estimation of matching positions on the sequences was obtained using the protein alignments. Differing amino acid residues were compared in the structure images. A distance search for potential spatially-connected residues was performed by the algorithm implemented in RasMol.

2.9 Web server

Due to the incredible growth in available kinesin sequences, a web interface and database are needed to store and update the sequence information. The following sections describe the features and construction of the kinesin web server.

2.9.1 MySQL database

A MySQL database (version '5.0.30-standard') with four tables was constructed. The table 'RefSeq_kinesin' stores the information of each kinesin sequence, including its RefSeq accession ID, RefSeq annotation name, organism, kinesin sub-family, protein sequence, and e-value of classification; The table 'Domain' stores the information of NCBI CDD classification including the accession ID of a sequence, starting and stopping of hits, the e-value, and classification result. The table 'Hmmer' is similar to the table 'Domain' and stores the corresponding information from a hidden Markov model classification. The table 'RefSeqalignment' is a table that stores all alignments, line for line.

An entity relationship diagram is shown in Figure 17.

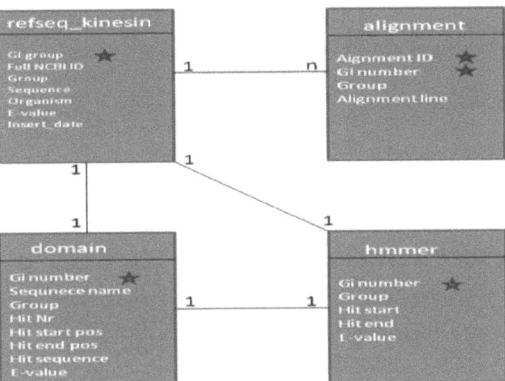

Figure 17. Entity model of the MySQL database. Four tables were created for saving general information on sequences and alignments. Each table is linked with a unique RefSeq accession ID: the gi number. Information stored in each table can be combined and used for generation of other results via external scripts.

2.9.2 Web interface

A web interface supported by LMU Biocenter web server can be accessed at www.bio.uni-muenchen.de/~liu/kinesin_new/. It provides up-to-date kinesin information, and many useful data analysis tools.

The main features of the web interface are listed as below:

- Table of kinesin sub-families
    - Number of kinesin sequences currently in the RefSeq database
    - Number of species that contain kinesins
    - Number of copies of kinesin in each sub-family in the corresponding species
- The distribution of kinesins in species
- Overview of kinesins in various species with the following information
    - NCBI access number and name
    - Sub-family classification with e-value
    - Motor domain start and end positions in the sequence
    - Protein sequence and amino acid usage distribution
    - Highlighted view of the motor domain
    - Link to protein conservation information
    - Link to alignment viewer
- Conservation information of a protein sequence in both absolute conservation and polar conservation in which amino acids are highlighted in different colors depending on their conservation score

- The classification engine can classify a given protein sequence that is provided in FASTA sequence format. Both NCBI CDD and hmmer classifier results are supported with e-values provided

- Known sequence motifs can be searched against the entire kinesin sequence database

- The motif search tool can be used to scan sequence motifs in kinesin sub-families or the kinesin super-family

- Alignments of kinesin sub-families or combinations of sub-families can be viewed in full length or in selected regions with options that allow conservation highlighting or mutational focusing

- The discriminating residue search tool calculates the discriminating residues by comparing alignments of two groups of sequences. The alignments can be user-defined or pre-calculated alignments of kinesin sub-families

- The resulting discriminating residues can be viewed on a 3D structure of the motor domain

- PDB entries of kinesin are assigned on corresponding sub-families. A conventional website with Jmol [106] was implemented to let users view the structure directly online

46

3 Results

3.1 Statistics of kinesin sequence data

3.1.1 Kinesin sequences in the RefSeq database

Until April.30th, 2009, 2,915 kinesin sequences were found and stored in the kinesin database. The length distribution of all kinesins is shown in Figure 18. 2,691 (92%) sequences have a length between 300 and 2,000 amino acids.

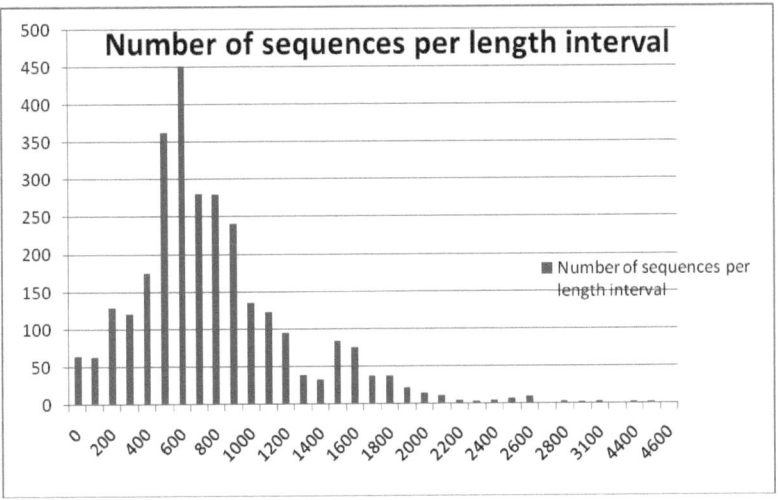

Figure 18. Length distribution of kinesin sequences in the RefSeq database (status on April.30th, 2009).

There are 132 sequences shorter than 300 amino acids, while the motor domains of kinesins have a length of about 325 amino acids in general and share a high degree of similarity among entire kinesin super-family. This means that these sequences are only partial kinesin sequences. Nevertheless, they were still identified as kinesins because they contain kinesin-specific motifs. For example, the shortest sequence in the kinesin dataset, the gi156362553 of *Nematostella vectensis*, has only 57 amino acids, although it could be clearly classified into the kinesin-3 family with an e-value of 3e-18 because it matches the kinesin-3 motor domain pattern 234-284 (Figure 19).

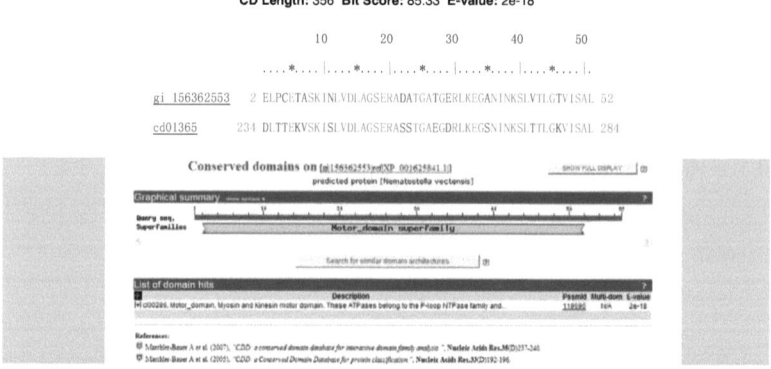

Figure 19. CDD conserved domain search of partial protein gi156362553. It hits domain ID cd01365, a kinesin-3. 50 out of 57 amino acids, from 2 to 52, match the domain pattern of cd01365, from 234 to 284, without a gap. E-value is 2e-18.

On the other hand, 92 sequences are longer than 2,000 amino acids. These sequences are predicted by gene prediction software. Due to uncertainties in these predictions, they contain not only typical kinesin sequences, but also extra sequence patterns.

For example, the gi189545928 of *Danio rerio* is 2,003 amino acids long. A conserved domain search shows that it belongs to the kinesin-2 sub-family with an e-value of 1e-32 (Figure 21). However, there is no other known domain or motif found in the rest of the sequence. By translating the mRNA of gi189545928 into protein sequence with highlighted start and stop codons (Figure 22), a long extra sequence appended to the end of the kinesin sequence is revealed.

Therefore, both partial and extra long sequences are excluded from the analysis (but still accessible via internet in the kinesin web server [82]), because they will make the alignments less reliable. For incomplete sequences, gaps will be inserted in the alignment to represent the missing parts of a sequence. In contrast, sequences with extra sequence patterns will cause long gap insertions in all other sequences or misalignments of certain parts of the sequences. All of these errors will lead to mistakes in the calculation of conservation or the reconstruction of ancestral states.

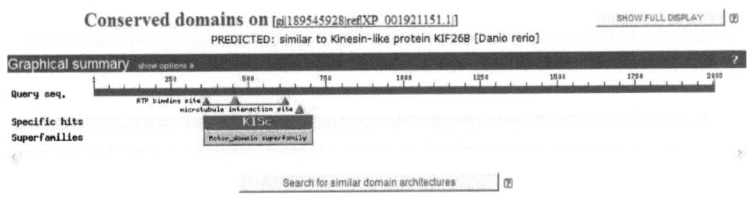

Figure 20. Conserved domain search of protein gi189545928. The graphical summary shows that the motor domain is offset by 356 amino acids from the N-terminus. In the rest of the sequence no known domain is found.

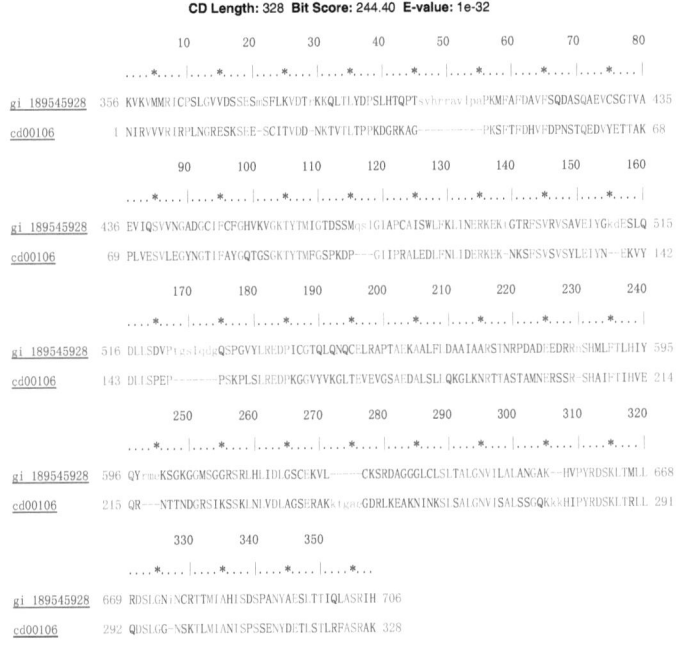

Figure 21. Alignment of gi189545928 with cd00106. Positions (356 to 706) match the entire cd00106 domain (1 to 328), which represents the kinesin motor domain.

Figure 22. gi189545928 with methionine and stop codons shown in bold. The sequence after the first stop codon is highlighted in pink. It indicates that the region predicted by the gene prediction software does not belong to a kinesin sequence.

3.1.2 Organisms which contain kinesins

The 2915 sequences of the RefSeq database are from 127 eukaryotic organisms, including 47 animals, 40 fungi, 14 apicomplexans, 8 green plants and 18 other organisms (Figure 24). Thus, the sequences represent a relatively complete set of organisms. They even include diplomonads, triplomonads and a choanoflagellate, the latter being at the base of the origin of all animals [83].

Figure 23 shows a schematic tree of the eukaryote kingdom. Except for two groups, the zygomycota (no sequences data in the database available yet) and the trematoda (only one organism included, *Schistosoma japonicum*), kinesins were found in every eukaryote lineage. This indicates that kinesin is an essential functional protein in almost every eukaryotic organism.

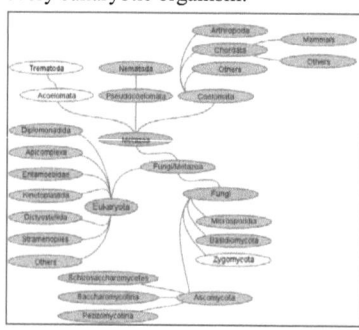

Figure 23. Map of kinesin distribution in the eukaryotic tree. A highlighted node indicates kinesins are found in at least one of its members.
http://www.ncbi.nlm.nih.gov/sutils/genom_tree.cgi

Figure 24. Tree of organisms created with the NCBI taxonomy browser. Numbers indicate how many organisms in each taxonomic group contain kinesins.

http://www.ncbi.nlm.nih.gov/Taxonomy/CommonTree/wwwcmt.cgi

Figure 25. Lineage tree of 127 organisms containing kinesins. Names of organisms are shown in bold. The entire tree is printed in 3 columns, from left to right.

3.2 Conservation analyses

3.2.1 Common conserved residues in the kinesin super-family

2,530 kinesin sequences were obtained from RefSeq release 33. Among these sequences, 2,350 kinesin sequences with a length of 300-2,000 amino acids were used to build a full-length clustalW alignment. In this dataset, the minimum length is 304 (115448625, kinesin-4, *Oryza sativa*), the maximum length is 1,990 (169852706, kinesin-4,*Coprinopsis cinerea okayama*). 2,342 (99.66%) of the sequences have either a complete or a partial motor domain (Table 6).

	>= 300	[125, 300)	< 125
Type of domain	complete	partial	Fragment
Number	2,241(95.36%)	101 (4.3%)	8 (0.34%)

Table 6. Number of sequences of different domain types. A motor domain is defined as complete when its length is greater than 300, partially complete when its length is between 125 and 300, and fragmentary when its length is less than 125. Only 8 (0.34%) sequences from 2,350 kinesin sequences are fragmentary.

To identify functionally and structurally important amino acids, three conservation calculation methods were applied for each position of the kinesin alignment: the absolute, polar and hydrophobic/hydrophilic conservation score. A position in the alignment is defined as conserved when one of the three scores is greater than 0.5. Overall, only three conserved residues (L345, M560, L713) were found outside of the head domain region. They are conserved as non-polar residues. All three of them are identified as conserved using the polar conservation calculation method. No 100% conserved residue in the whole kinesin alignment has been identified. In contrast, in the head domain region, 101 conserved positions were identified by the absolute conservation method, 187 conserved positions were identified by the polar conservation method, and 191 conserved positions were identified by the hydrophobic/hydrophilic conservation method. All 101 conserved residues are also conserved according to the polar and hydrophobic/hydrophilic method. 18 residues were only polar conserved, 22 residues were only hydrophobic conserved (Table 7). Thus in total 209 conserved residues are identified in the head domain. Among them, 3 residues are found outside of the motor domain; 49 residues are located in the loop regions; the rest (157 residues) are located in secondary structure elements (Figure 26).

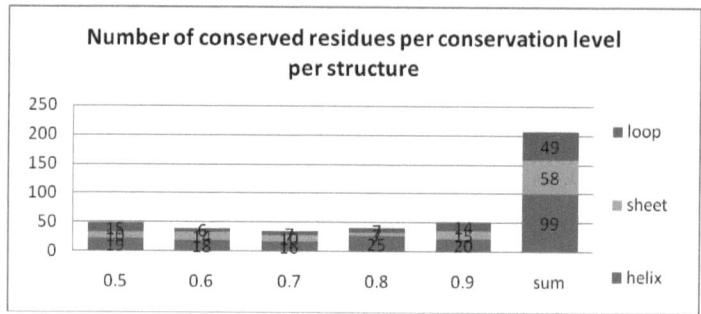

Figure 26. Number of conserved residues at different conservation levels and their locations in the motor domain.

15 of 18 polar conserved residues are located in the secondary structure elements (α helix or β sheet). Their conservation levels are all less than 0.7, just above the conservation threshold. 17 of them are either non-polar or uncharged residues.

Only one glutamic acid (E316) on α6 is polar and conserved. Structure analysis reveals that E316 shows hydrophilic interactions with Q89 on the p-loop and S239, E240, K241 on switch II. Thus, E316 could be important for the spatial positioning of the p-loop, switch II, and α6.

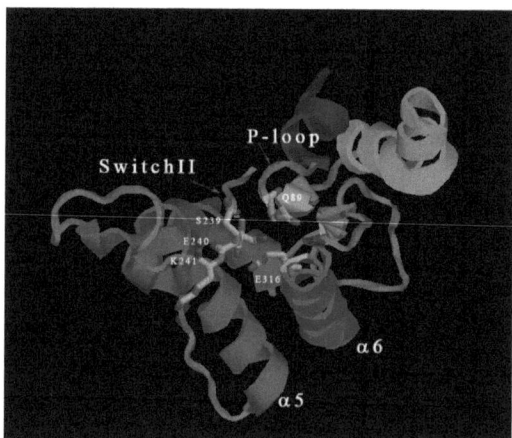

Figure 27. Polar conserved residue E316 in contact with Q89 from the p-loop and S239, E240, K241 from switch II, shown here in a partial secondary structure of kinesin-1 (1GOJ).

Of the 22 hydrophobic/hydrophilic conserved residues, only V14 is hydrophobic and conserved. The others are either hydrophilic or neutral conserved. Six residues are located in the loop regions: one in loop3, two in Switch I loop, one in Switch II loop, one in loop13 and one in the neck linker. N202 and E203 in the switch I, R251 in switch II and E347 in the neck are hydrophilic conserved.

Previous studies of several kinesin motor structures have shown that switch I is a functionally important motif, responsible for conformational changes of the motor domain in response to ATP hydrolysis and the release of Mg^{2+} and ADP [84]. The conservation analysis shows that switch I is over 50% hydrophilic conserved in the entire kinesin super-family. This underscores the importance of the switch I motif in kinesins. It is possible that the switch I motif is inherited from the ancestor of kinesins. On the other hand, the R251 residue is located in the turning point of switch II and has no direct interaction with any functional residues. Therefore, it should be a structurally important residue as well.

a)

Positions	Residues	Polarity	Value	Location	positions	Residues	polarity	Value	Location
11	V	Non-polar	0.58	β1	218	Q	Uncharged	0.57	β6
34	V	Non-polar	0.50	L3	232	N	Uncharged	0.61	β7
69	V	Non-polar	0.53	α1	306	T	Uncharged	0.69	β8
76	Y	uncharged	0.61	α1	313	N	Uncharged	0.51	α6
99	M	Non-polar	0.51	α2a	316	E	Polar	0.55	α6
167	V	Non-polar	0.51	β5c	319	N	Uncharged	0.54	α6
172	L	Non-polar	0.54	β5loop	345	I	Non-polar	0.51	Neck
180	V	Non-polar	0.67	α3	560	L	Non-polar	0.51	–
213	T	Uncharged	0.72	β6	713	L	Non-polar	0.50	–

b)

Positions	Residues	hydrophobic	value	Location	Positions	Residues	hydrophobic	value	Location
14	V	Yes	0.58	β1	198	A	Neutral	0.71	α3
39	D	No	0.51	β2	200	N	no	0.50	α3a
56	P	Neutral	0.50	L3	202	N	No	0.50	Switch I
70	E	No	0.55	α1	203	E	No	0.52	Switch I
75	G	neutral	0.60	α1	229	G	neutral	0.75	β7
121	E	No	0.56	α2	251	R	No	0.50	Switch II
124	E	No	0.52	α2	257	N	No	0.61	α4
133	Y	Neutral	0.50	β4	260	K	No	0.53	α4
162	P	Neutral	0.51	β5	292	Q	No	0.80	α5
188	E	No	0.51	α3	310	A	Neutral	0.72	L13
195	T	neutral	0.66	α3	347	E	No	0.52	Neck

Table 7. Table of conserved residues. Position numbers refer to the kinesin-1 sequence gi164422752 of *Neurospora crassa*. Residues are consensus residues obtained from the kinesin alignment, indicating the most conserved residue at certain positions. a) Conserved residues only detected by the polar conservation method. b) Conserved residues only detected by the hydrophobic/hydrophilic conservation method.

Matching the conserved residues on the kinesin crystal structure could give us valuable insights into the conservation of the structure of the motor domain. In particular, conservation regions of known functional motifs and structural elements will be revealed at a glance.

After mapping the conserved residues onto the secondary structure of the kinesin motor domain, it is clear that most of the important regions, either functional motifs or secondary structural elements, are almost entirely comprised of conserved residues (Figure 28, Figure 29). For example, the P-loop in the kinesin-1 motor of *Neurospora carassa* consists of five residues. Four of them are the same in over 92% of all kinesin sequences. A similar degree of conservation can also be found in the switch II motif and the ATP binding pocket (Table 8).

Name	Position	Consensus
P-loop	88-92	$G^{0.97}Q^{0.78}T^{0.95}G^{0.94}[Sa]^{0.92}$
Switch I	202-203	$N^{0.5}E^{0.52}$
Switch II	236-241	$L^{0.98}A^{0.97}G^{0.97}S^{0.85}E^{0.91}[Rk]^{0.91}$
ATP binding pocket	13,15, 89-97, 235	$R^{0.77}, R^{0.81},$ $Q^{0.76}T^{0.95}G^{0.94}[Sa]^{0.92}G^{0.97}K^{0.96}[Ts]^{0.96}Y^{0.75}T^{0.96},$ $D^{0.98}$

Table 8. Conservation of functionally important motifs in the kinesin super-family. A lower case letter indicates that the corresponding residue in kinesin-1 of *Neurospora carassa* is not same as the most conserved residue at this position in other kinesins. For example, $[Sa]^{0.92}$ indicates that 92% of kinesin sequences use Serine at position 92, while an Alanine occupies this position in kinesin-1 of *Neurospora carassa*.

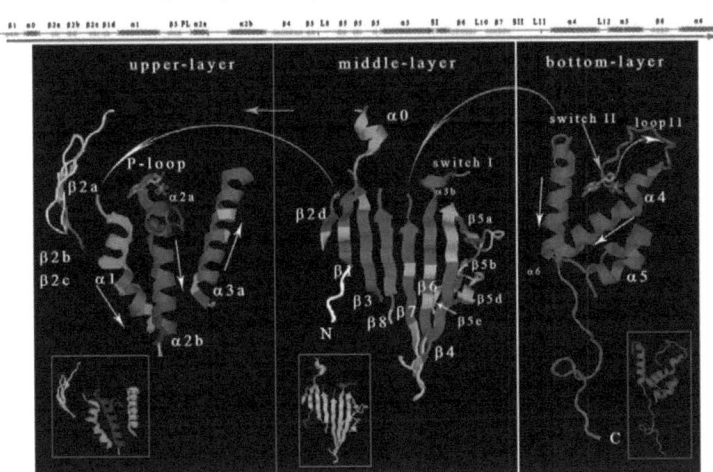

Figure 28. Overview of the structure of a kinesin-1 motor domain (1GOJ) dissected into the three layers of secondary structure elements. A linear structure is shown on the top of the image (green for β-sheets and blue for α-helices). Colored secondary structure elements are shown in the insets. Conserved residues are colored in red. The remaining polar and hydrophobic/hydrophilic conserved residues are colored in gray.

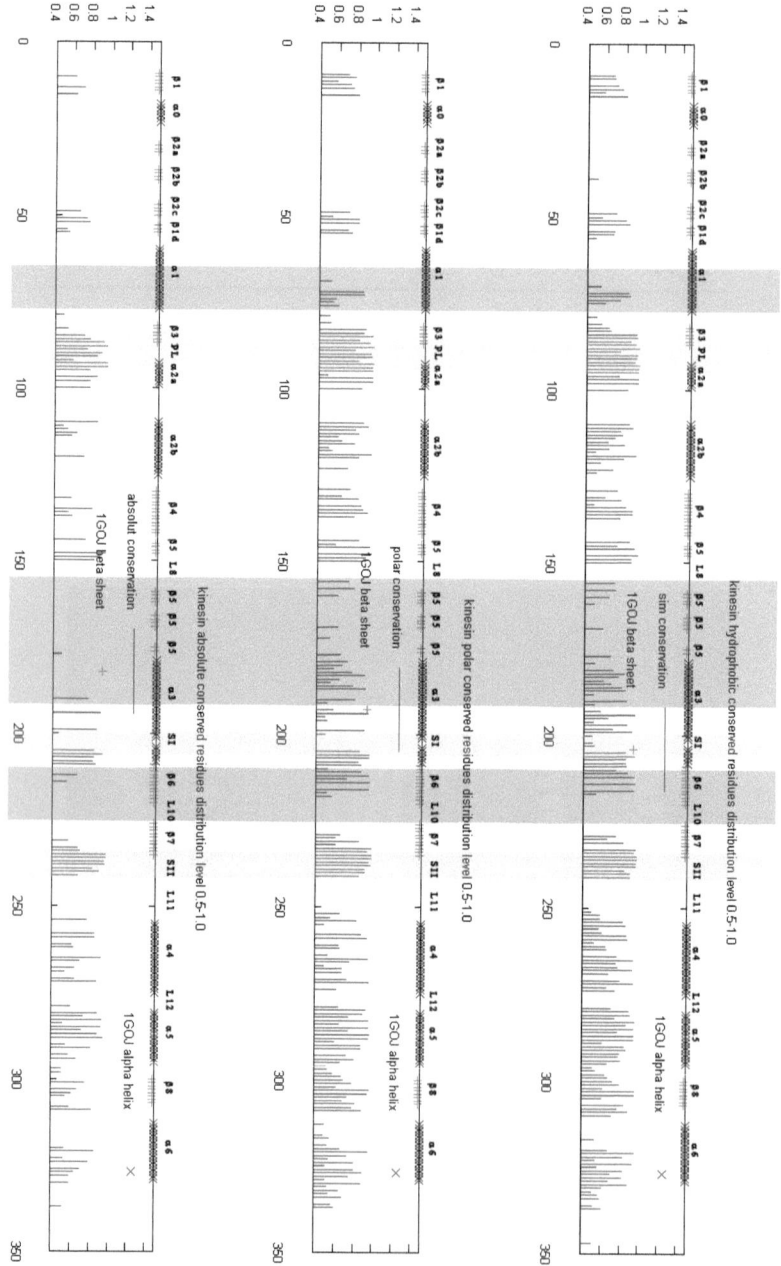

Figure 29. Global conservation of the kinesin super family. The graph shows the hydrophobic, polar, and absolute conservation of the motor domain, respectively. Secondary structure of 1GOJ (*Neurospora crassa* kinesin-1) is shown with colored dots, green dots indicate β sheets, blue dots indicate α helices. Conservation scores of amino acids are indicated with red impulses, less than 50% conserved residues are not shown. The regions highlighted in yellow indicates the p-loop, switch-I and switch-II motifs. In the regions highlighted in gray no absolute conservation but polar and hydrophobic/hydrophilic conservation has been found.

3.2.2 Common motifs of kinesins

Based on 106 complete kinesin motor domain sequences, previous studies have identified six highly conserved motifs (named A-F) and two additional motifs (named X,Y) [http://www.proweb.org/kinesin], which are found in most kinesins. All of them have been found by the motifscan program (see methods). Many positions in the motifs have now been updated and refined: Some residues previously believed to be conserved cannot be confirmed in the large kinesin dataset. For example, the motif X [IxVxCRCRPxxxxE] is updated to [IxVxVRxRP], where the Glutamic acid at the end of the motif X and the cysteine at position 8 are not conserved any more. At position 5, valine replaced cysteine as the most conserved residue.

Eight more motifs have been noted in addition, because their conservation levels are all above the average conservation level (31.56%) of the entire kinesin family (Table 9). These motifs represent important structure elements in the motor domain, such as loop3 linked with α1, loop8, α6, the end of the motor domain, and the neck linker. These 16 motifs well describe the kinesin motor domain structure (Figure 30).

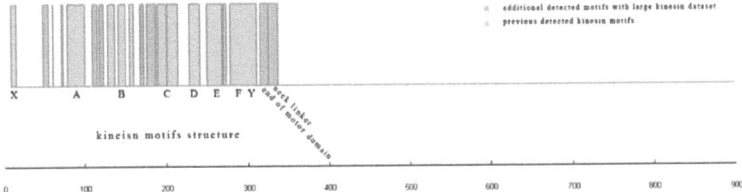

Figure 30. Motif structure of the kinesin super-family. 16 motifs have been detected in the large kinesin dataset with 2,350 sequences. In addition to the 8 motifs detected earlier (in green), 8 additional motifs have been identified (in light blue). All are located in the head domain region.

	Motif	Con.	Positions	Previous detected motifs	name	Structure
1	IxVxVRxRP	0.43	9--15	IxVxCRCRPxxxxE	X	β1
2	FxFDxVF	0.53	49--55			L3
3	Q	0.42	61--61			α1
4	VL	0.33	72--73	VxxVLxGYNCCIFAYGQTGSGKTYTMxG	A	α1
	GYNxTIFAYGxQTGSGKT	0.65	79--95			L4-β3-ploop-
	YTMxG	0.61	96--100			α2a
5	GIIPRA	0.58	110--115			α2b
6	LFxxI	0.37	119--123			α2b
7	VxxSYLEIYN	0.46	129--137	FxVxVSYFEIYNExIRDLL	B	β4
	YxExIRDLL	0.54	142--150			L7-β5a-L8a
	LxVxE	0.31	156--160			L8b-β5b
8	VxxL	0.30	169--172			L8c
9	VxSxxExxxLL	0.30	177--187			L8d-α3a
10	GNxNRxVAATxxN	0.45	190--200	RxVAxTxMNEHSSRSHAIFxI	C	α3a
	NxxSSR	0.57	202--207			switchI-α3b
	SHAIFTI	0.60	208--214			β6
11	GKLxLVDLAGSER	0.72	229--241	GKLxLVDLAGSER	D	β7-switchII
12	RLKEAxxINKSL	0.53	251--262	EAQNINQSLSCLGxCIxAL	E	L11-α4
	LxALGNVI	0.50	263--269			α4
	ALxD	0.54	271--274			α4
13	HIPYR	0.73	279--283	HIPYRDSKLTxLLQDSL	F	L12-
	DSKLTR	0.75	284--289			α5-
	LLQDSLGG	0.56	290--297			L13
14	GNSKTxMI	0.50	297--304	KTxMIACCSP	Y	L13-β8
	AxISPA	0.50	305--310			β8-L14
15	ETxxTLRYA	0.51	316--324			α6
16	RAKxIKNKxxxN	0.44	326--337			α6-neck link

Table 9. Motifs detected by motifscan, compared with the previously noted eight common motifs in the kinesin super-family. Motifs are ordered by their position in the primary structure. The reference sequence for position mapping is kinesin-1 of *Neurospora crassa* (gi164422752).

3.2.2.1 ATP-binding pocket

The residues with a distance of less than 5.0 Ångstroms to the ADP molecule form a pocket. It is comprised of residues 13, 15, 16, the p-loop 89-92, the beginning of helix α2 93-97, and residue 235 of β7. Among these, residues 91-96 are less than 3.5 Ångstroms away from the nucleotide. The average conservation of these residues is over 81.5%. For example, the positively charged K94, which is normally used to bind the ATP and replace Mg^{2+}, is 97% conserved in the entire kinesin super-family.

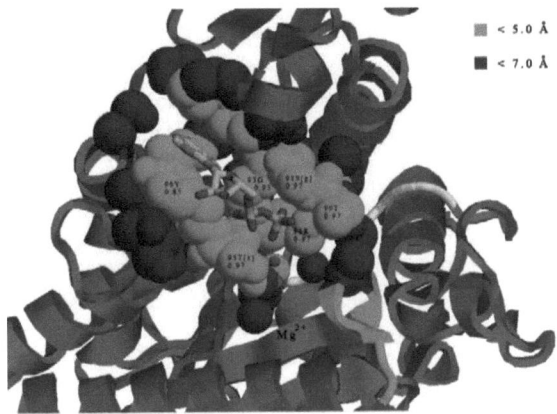

Figure 31. Nucleotide binding pocket with conservation value. The ADP (stick model) is located inside the pocket formed from amino acid residues highlighted in green. Blue-colored residues have a distance of between 5.0 to 7.0 Ångstroms from ADP. Green-labeled residues and ADP are 3.5 Ångstroms apart.

3.2.2.2 Switch I and switch II motifs

The switch I and switch II motifs are thought to be important for the conformational changes of the motor domain during ATP hydrolysis. Switch I is formed from residues 202-207. Except for the position 204, which is less than 50% conserved, the other residues are, on average, 75% conserved. Furthermore, switch II, which comprises residues 236-241, is over 93% conserved in all kinesin sequences (Figure 32).

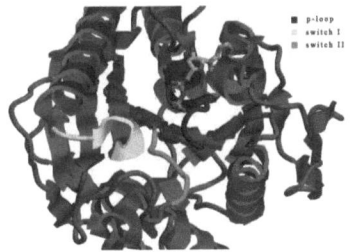

Figure 32. Highly conserved functional motifs near the ATP binding site: p-loop in blue, switch I in yellow, switch II in green.

Because the determination of the degree of conservation of residues is based on an analysis of the entire kinesin super-family (over 2,300 sequences in 127 organisms), the high incidence of conserved residues in these regions demonstrates that these regions are of general importance for all kinesins. This leads to other interesting questions. What is the importance of other residues in these motifs? How have they evolved? To answer these questions, it is necessary to investigate the phylogeny of kinesins. In general, a protein family consists of sub-families, which evolved from a common ancestor. The relatedness of sub-families can be visualized in a phylogenetic tree. Members from the same sub-family are closely related and represent a clade (sub-tree) within the phylogenetic tree. With an accurate phylogenetic tree of a protein family, one could easily map the important residues or motifs onto the tree and trace their evolutionary history.

3.3 Phylogenetic analysis and kinesin classification

2,530 kinesin sequences were used to construct a phylogenetic tree of the entire super-family. The tree was rooted using the midpoint rooting method [85]. A comparison with the outgroup method has shown that the more consistent the outgroup root is, the more accurate the midpoint rooting method appears to be. This means that the midpoint rooting method should reveal a similar phylogeny compared to using multiple outgroup taxa to root the tree. For a large kinesin dataset with more than 2,500 protein sequences, the diversity of sequences is great. In this case, including extra multiple outgroup sequences would make it harder to align the sequences correctly, affecting the accuracy of the phylogeny. Therefore, the midpoint rooting method was a reasonable alternative to root the tree.

To infer a phylogenetic tree with such a large dataset is a great computational challenge. The Bayesian method mrbayes [86] has failed to analyze the dataset within a reasonable time frame. The neighborhood joining method was able to calculate the phylogenetic tree, however, due to the high diversity of the sequences analyzed, many sequences have been placed into wrong sub-trees, and the tree was thus not reliable.

RaxML, a fast maximum likelihood-based program, was successfully used to analyze the alignment and to construct a kinesin tree. The quality of a phylogenetic tree largely depends on how well the sequences from different families are grouped. This means, that when one uses two sets of protein sequences of two different protein sub-families to build a phylogenetic tree, a correct tree should contain two clearly separated clades, each of which should be formed by one group of sequences. In other words, the rate of how many sequences were correctly placed in the tree could be used as a measure of the quality of the tree. To do so, the sequences used for tree building were first classified by CDD classification and assigned a sub-family number. After the tree was built, the quality of the tree was determined by counting the number of sequences with same group number in one clade.

3.3.1 Classification based on the Conserved Domain Database

Sequences were classified in the first step using CDD classification. Each sequence was assigned to one unique sub-family. The statistics are shown in Table 10. More than 100 kinesins were detected in some of the previously defined kinesin groups, such as kinesin sub-families 1, 2, 3, 4, 5, 6, 7, 8, 13, and 14. Three other groups, kinesin-9, -10 and -11, were not clearly defined before due to a lack of sequence data. In the present large kinesin dataset, there are more than 50 sequences assigned to each of these groups. This is a sufficiently large number to build clear clades in the tree.

In the remaining class, kinesin-12, a total of 262 sequences were grouped. This class is defined as the "orphan" group in the CDD classification scheme. They are called orphans because members of this group did not have similar sequence profile and could not be defined as a unique kinesin sub-family. Some of the sequences in this family could be classified with the help of the phylogenetic tree later. Orphans that occur inside of one clearly defined clades of other kinesin sub-families are defined as false negatives. These sequences are then assigned with the same class name as the other members in that clade.

3.3.2 Phylogenetic tree of the kinesin super family

The number of sequences of each kinesin sub-family predicted by the CDD classification was compared with the number of sequences grouped together in a tree clade in Table 10.

Group	CDD	Subtree	K12	%	Group	CDD	Subtree	K12	%
K1 K1.1	183	126/126 28/31	3	85%	K8	208	199/200	0	96%

K2	235	221/226	0	96%	K9	76	75/78	2	98%
K3	373	348/353	3	93%	K10	49	46/50	4	93%
K4 K4.1	232	163/165 37/37	0	91%	K11	53	51/59	3	96%
K5	111	111/112	1	100%	K12	262	---------		
K6	111	106/137	31	96%	K13	226	225/229	4	99%
K7 K7.1	154	93/107 40/40	9	87%	K14	257	225/227	2	88%

Table 10. CDD classification vs phylogeny-based classification. Column CDD shows the number of sequences of a kinesin group predicted by CDD classification. Column Subtree shows the information of a clear clade found in the tree. For example, 221/226 for kinesin-2 (K2) means that there are 226 sequences included in the clade, 221 of which belong to the kinesin-2 group according to the CDD classification. Column K12 shows how many sequences in the clade were from the kinesin-12 group; they are identified as false negatives within the corresponding kinesin group. For example, 31 sequences in the kinesin-6 clade were predicted as kinesin-12 by CDD classification. According to the phylogenetic tree, these sequences should be members of kinesin-6. Column % indicates the accuracy rate of a clade, which represents a kinesin sub-family in the tree. It was calculated by the number of sequences of a kinesin group included in a clade divided by the number of sequences of the kinesin group.

Thirteen of fourteen standardized kinesin sub-families were clearly detectable in the tree (Figure 33). The classification based on the tree was highly consistent with the CDD classification result (Table 10).

The accuracy rates of clades in the tree were calculated. They show that for each kinesin group, more than 85% of its sequences predicted by CDD classification were grouped together to form a clade in the phylogenetic tree. The kinesin-5 clade even includes 100% of predicted kinesin-5 sequences. Overall, 2,156 of 2,530 sequences were clearly grouped into 16 clades (including 3 small clades: kinesin-1.1, kinesin-4.1 and kinesin-7.1). This makes for an overall accuracy of the tree of 85.7%.

There are 62 orphan sequences found within these 13 sub-clades. They have been reassigned to the corresponding kinesin groups.

Besides these 13 large groups, there are several small clades (colored grey in Figure 33) built from mixed members of the orphan group kinesin-12 and various other kinesin groups. For example, the clade labled "orphan" includes 92 sequences, 67 of which were classified as kinesin-12; the rest are from kinesin groups 3, 4, 8, and 14. The clade adjacent to the kinesin-6 contains 87 sequences, 55 of which were predicted as kinesin-12, and the rest are from kinesin groups 1, 3, 4, 7, 8, and 14.

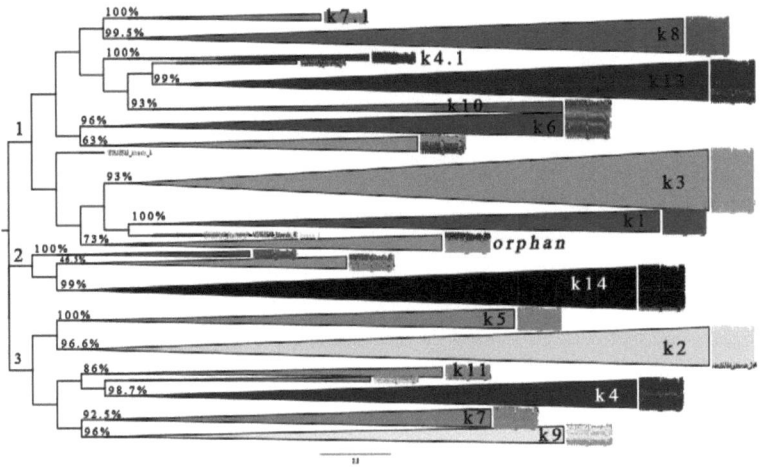

Figure 33. Phylogenetic tree of 2530 kinesin sequences, built using the maximum likelihood method implemented in the RaxML and rooted using the midpoint rooting method. 13 kinesin sub-families are clearly detectable.

According to this phylogeny, all the kinesin sub-families can be grouped into 3 larger groups: kinesin families 1, 3, 6, 8, 10, 13 in clade 1; kinesin families 2, 4, 5, 7, 9, 11 in clade 3; and kinesin-14 alone in clade 2 (Figure 33). This could indicate the first duplication in the evolutionary history of kinesins. After that, 2 or 3 further duplications happened in two of the clades, leading up to the present-day kinesin super-family. For example, in clade 1, after the second duplication, one of the duplication products has evolved into the main kinesin-3 family. Then a third duplication occurred and the kinesin-1 class evolved from one of the duplication products of kinesin-3. The other branch of the second duplication underwent further duplication events, leading to the formation of the kinesin classes 6, 8, 10 and 13. In clade 3, three additional duplications can be observed. The kinesin classes 2, 4, 5, 7, 9 and 11 have evolved from these duplication events. In clade 2, after two duplications, no new kinesin sub-family has formed and kinesin-14 has become the major member in this clade.

Expanded views of the kinesin sub-family trees from kinesin-1 to kinesin-14 are accessible online at [www.bio.uni-muenchen.de/~liu/kinesin_new/phylogeny.php].

3.3.3 Evolution of kinesins

With the classification of kinesins in hand, it is easy to determine how many kinesin sub-families an organism has. In combination with a taxonomic tree, the evolutionary history of kinesins among species can be uncovered. Figure 34 shows the mapping of the kinesin groups of each species on the standard taxonomy tree provided by the NCBI.

It is noteworthy that all kinesin families are represented in *Trichoplax adhaerens* and *Nematostella vectensis*, the roots of the metazoan group in this dataset, and in most of the mammalian species. Only kinesin-7 is not found in *Bos taurus*, indicating the loss of an entire group in this group of animals during evolution. The full compelment of kinesin super-family members is also found in the sea urchin *Strongylocentrotus purpuratus*. But in other species from the deuterostomia group, the loss of one or more kinesin sub-families is rather widespread. For example, there is no kinesin-8 in *Ciona intestinalis*, there is no kinesin-7 or kinesin-11 found in *Danio rerio*, and there is no kinesin-9 in *Xenopus*. Kinesin-10 and kinesin-11 are lost in the entire neuroptera group. Except for *Apis mellifera*, no kinesin-9 is found in other insects.

On the other hand, representatives of kinesin-2, kinesin-9 and kinesin-11 are not present in the current fungal dataset. Kinesin-13 is often lost in members of the basidiomycota group. Kinesin-6 is not found in any species from the ascomycota group. Moreover, kinesin-10 and kinesin-13 are lost in both saccharomycetales and schizosaccharomyces.

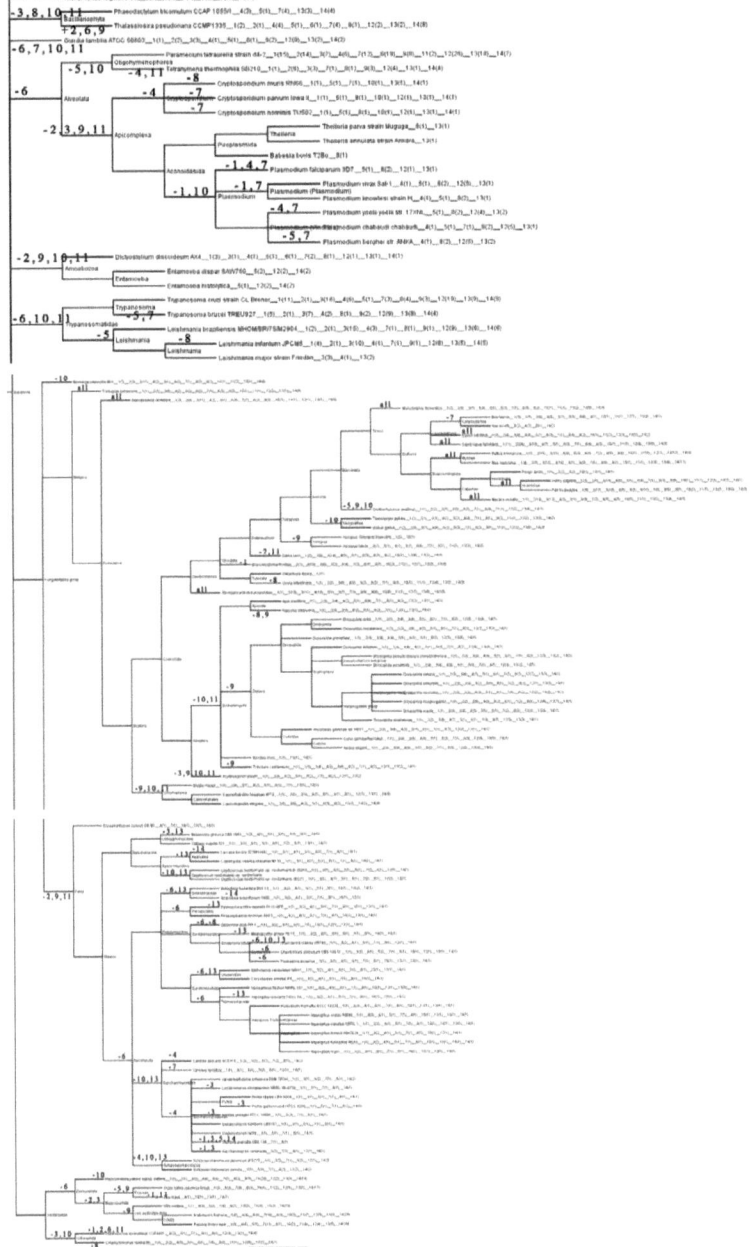

Figure 34. The distribution of kinesins in organisms, together with the taxonomic tree, give an overview of the evolution of kinesins. The expanded figure can be viewed online at
[http://www.bio.uni-muenchen.de/~liu/kinesin_new/images/grp.org.png]

groups. Kinesin-4 is only found in *Yarrowia lipolytica* but not other species from this group. In the other subphylum of the ascomycota, the pezizomycotina, kinesin-6 is lost in all species except *Magnaporthe grisea*. And it seems that kinesin-13 is tending to disappear in most of the species.

In the early eukaryotes, such as *Giardia lamblia,* amoebazoa, alveolata etc., nine or ten of the kinesin sub-families are represented. This demonstrates that most kinesin sub-families were already present early in the evolution of extant eukaryotes.

Most of the organisms contain representatives of all three main kinesin clades in the tree except the aconoidasida group and *Trichomonas vaginalis*, which do not have kinesin-14. Two adjacent kinesin sub-families never get lost in one species at the same time, indicating the independence of the three main clades and the high similarities of adjacent kinesin sub-families. It suggests that kinesins of each of the main clades may be required for different cellular functions. On the other hand, adjacent families likely serve similar functions in the cell, so that only one of them is retained in some species during evolution.

3.4 Kinesin-1

3.4.1 Sequence data

The kinesin-1 sub-family is known as conventional kinesin. It is the best-studied kinesin class. In the 2,530 kinesin sequences, 183 sequences were predicted as kinesin-1 by the CDD classification. 126 of these were found in one single clade in the phylogenetic tree of the kinesin super-family. These 126 sequences have been identified as true members of the kinesin-1 sub-family. The analyses of kinesin-1 were based on this dataset.

These sequences are from 86 distinct eukaryote species, including 40 animals, 31 fungi, 4 green plants and 11 other simple organisms, such as kinetoplastids, ciliates, diplomonads and choanoflagellates (Figure 35).

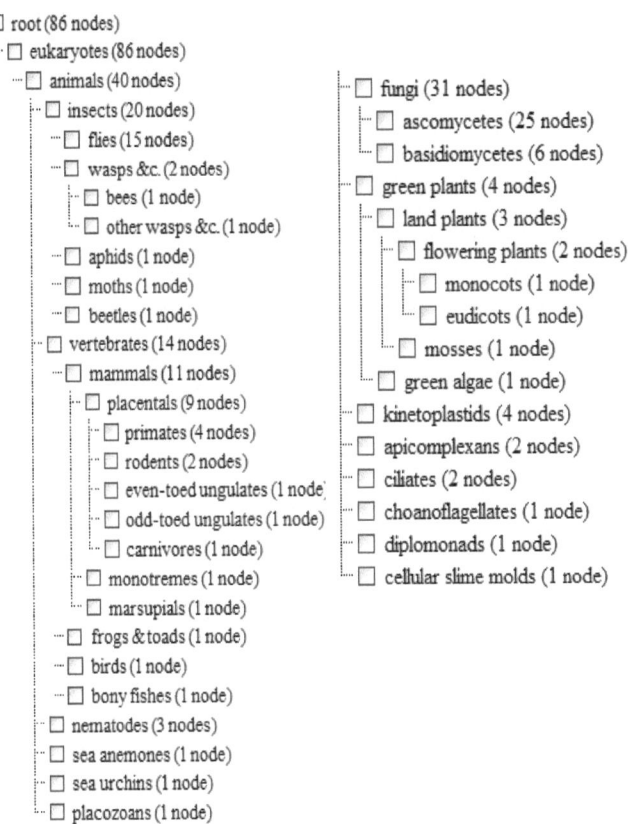

Figure 35. Taxa distribution tree of kinesin-1 based on 126 true kinesin-1 sequences. Ambiguous sequences were excluded from the analysis.

The lengths of these sequences vary from 285 aa to 1,380 aa. 15 sequences (11%) are shorter than 700 amino acids. They are defined as partial sequences. Two sequences are shorter than 300 amino acids. They are identified as fragments. The partial and fragmental sequences were included in the analysis. All of these sequences contain a nearly complete motor domain. Moreover, some of these sequences represent some unique species. Exclusion of these sequences means to remove the corresponding species from the dataset, which will limit the sampling of taxa. In fact, phylogenetic tests have shown that including these partial and fragment sequences did not affect the quality of the phylogeny.

3.4.2 Conservation of kinesin-1

Conservation analysis of the kinesin-1 dataset has shown that there are 204 residues that are over 90% conserved. All of them are found within the motor domain and neck region. Except for loops, α0 and β2a-c, every secondary structure element is found to be highly conserved in the motor domain. The less conserved structure elements are found to have a 60% conservation level (Figure 36). The crystal structure of the *Neurospora crassa* kinesin-1 (1GOJ) motor domain shows 355 amino acids consisting of 13 helices, (83 amino acids); 21 strands, (129 amino acids); and loops (143 amino acids). At a 60% conservation level, 691 amino acids are found conserved in the kinesin-1 group in total (Figure 36). 314 of them locate in the head domain (the crystal structure), which means that over 88% of the head domain of kinesin-1 is composed of conserved residues (Table 11). Of these 314 residues, 21 residues are even 100% conserved.

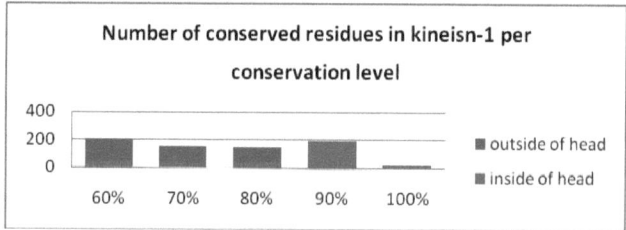

Table 11. Numbers of conserved residues in the kinesin-1 group found at different conservation levels are shown in boxes. According to the kinesin-1 crystal structure (1GOJ), the first 355 residues are considered to comprise the head domain. Residues with a position > 355 in the sequence are located outside the head domain.

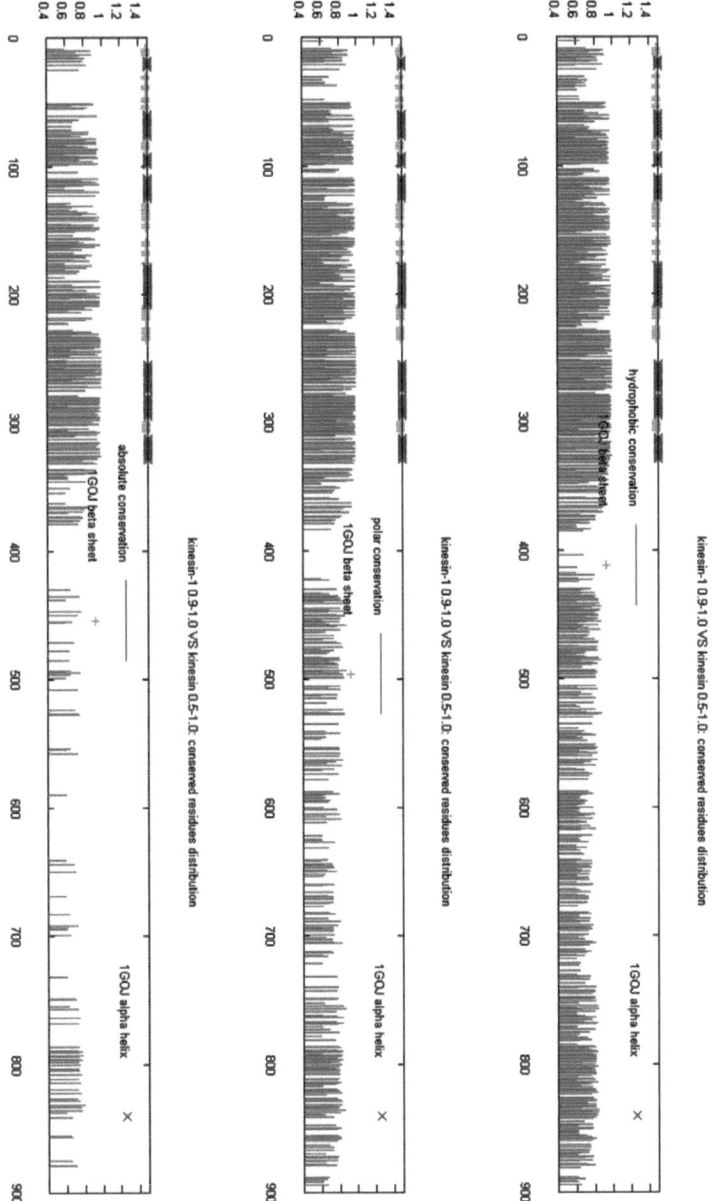

Figure 36. Map of amino acid residues with a conservation level of over 60% in the kinesin-1 group.

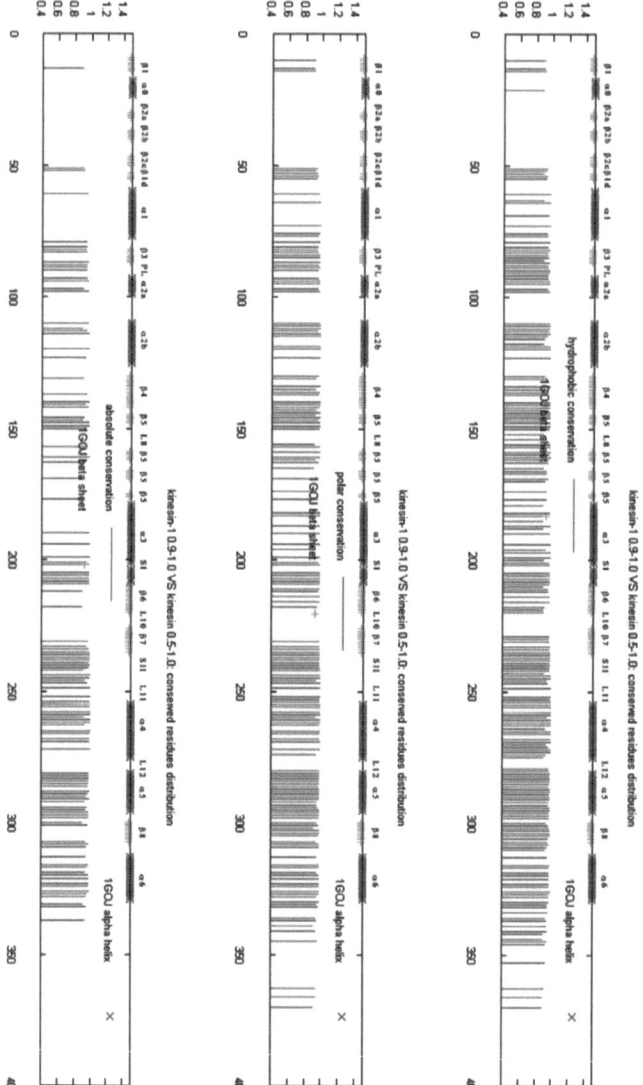

Figure 37. Residues in the kinesin-1 class that are over 90% conserved under three different conservation criteria, compared with *Neurospora crassa* kinesin-1 secondary structure.

Of the 21 residues that are 100% conserved, 10 are 100% absolute conserved, 9 are 100% hydrophobic conserved and the other 2 are 100% non-polar conserved. They are all found in the microtubule-binding-associated structure elements loop-11 (7 residues), α4 (13 residues), and α6 (1 residue) (Table 12).

Structure	Position*Residue (100% conserved criteria)
Loop 11	242V(non-polar), 244K(hydrophilic, polar charged, absolute), 245T(neutral, polar uncharged), 247A(non-polar), 249G(neutral, non-polar, absolute), 252L(hydrophobic, non-polar, absolute), 254E(hydrophilic)
α4	255A (neutral, non-polar, absolute), 256L(hydrophilic, polar charged), 258I(hydrophobic, non-polar, absolute), 259N(hydrophilic, polar uncharged, absolute), 261S(neutral, polar uncharged, absolute), 262L(hydrophobic, non-polar, absolute), 263S(neutral), 265L(hydrophobic, non-polar, absolute), 266G(neutral), 268V(hydrophobic), 269I(hydrophobic, non-polar), 272L(hydrophobic, non-polar, absolute), 273T(neutral)
α6	326R(hydrophilic)

Table 12. List of 100% conserved residues found in the kinesin-1 group. Residues are mapped with their corresponding structure and shown in the form position*residue (100% conserved criteria). For example, 244K (hydrophilic, polar charged, absolute) indicates the lysine at position 244 is 100% conserved under all 3 conservation criteria: hydrophilic, polar charged and absolute. Absolute conservation means that all observed kinesin-1 sequences have the same residue at this position.

These three secondary structure elements are believed to interact directly with the microtubule [87-91]. The 100% conservation observed in these regions indicates that loop-11 and α4 are strongly under positive selection. The α4 helix comprises 20 residues in the *Neurospora crassa* kinesin-1 sequence. 13 of them are found 100% conserved, while the rest are conserved at 80% -- 90%. The high conservation level of the α4 helix could be due to the need to interact with the microtubule surface. As the primary and secondary structure of microtubules is highly conserved at the interaction site, so are the corresponding interaction sites in the kinesin motor domain. In other words, these residues have to be 100% conserved in order to create the tightest binding with the microtubule.

In the loop-11 region, 7 residues are found to be 100% conserved. Of these, three are 100% absolute conserved. Figure 38 shows these three residues within loop-11 in the 3D structure. It shows that the residues from 244 to 252 of loop-11 form a 'Ω' like horseshoe shape. The three residues locate at the beginning (lysine244), the middle (glycine249) and the end (leucine252) of the horseshoe, respectively. This shape could be specific for kinesin-1, because these residues are not conserved any more in other kinesin families. Only the LEU252 is observed to be conserved at a 60% level. It suggests that LEU252 is functionally important as part of the microtubule binding site. Figure 38 also shows that the side chain of LEU252 interacts with the side chain of LYS244. It ensures to form and stabilize the horseshoe shape.

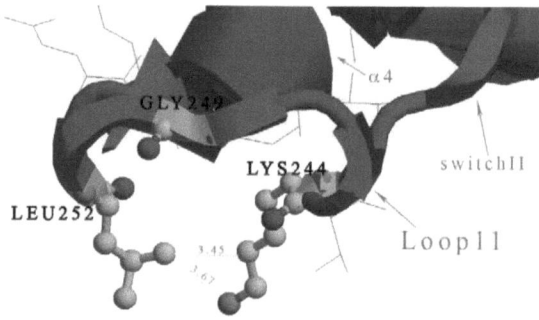

Figure 38. Structure of loop11 with three 100% absolute conserved residues: LYS244, GLY249, LEU252.

3.4.3 Motif structure of the kinesin-1 sub-family

Motif scans has shown that the overall conservation level of the kinesin-1 family is 50%. 27 motifs were detected in total. The motif structure of the kinesin-1 family is shown in Figure 39. Detailed information of the motifs is listed in Table 13.

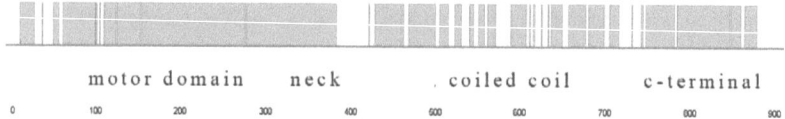

Figure 39. Motif structure of the kinesin-1 sub-family.

Motif	Con.	Position	Structure
ExxIKVxCRFR	0.62	9—15	β1
PLNxxExxxG	0.51	16—25	α0
YxFDRVF	0.71	49—55	L3
QExVYxxxAKxIVxDVLxGYNGTIFAYGQTSSGKTHTME	0.72	61—99	α1-L4-β3-ploop-α2a
GIIPRIVxDIFxxIY	0.73	110—124	α2b
MxENLEFHIKVSYxEIYMDKIRDLLDVxKxNLxVHEDKNRVPYVKGxTERFVSSPEEVxxVIxEGKxNRHVAVTNMNEHSSRSHSIFLINVKQENxETxxKLSGKLYLVDLAGSEK	0.74	126—241	L6-β4-L7-β5-L8-α3a-switchI-α3b-β6-L10-β7-switchII
VSKTGAEGxVLDEAKNINKSLSALGNVISALADG	0.85	242—275	L11-α4
THVPYRDSKLTRILQESLGGNxRTTxxICCSPSSxN	0.85	278—313	α5-L13-β8-L14
ExETKSTLxFGxRAKTIKNxxxVNxELTAEEWKxxYEKEKEKxxxLxxxIxxLExELxRWRxGExVPxxE	0.61	314—383	α6-neck linker
ExxxLYxQLDxKDxEINQxSQxxExLKxQxxxQEELxA	0.50	429—459	No structure available, predicted coiled coil region
QxELxxLQxENxxxKxEVKEVLQALEELAVNY	0.52	469—500	
DxKSQExxxK	0.50	506—515	
LxxELxxK	0.50	524—531	
ELxxL	0.47	541—545	
QxKRxxE	0.52	552—558	
LxxDLxExG	0.46	563—571	
ExFTxARLxxSKxKxExK	0.50	590—607	
LxISQHEAxxxSL	0.52	638—650	
ExKxRxLEExxDxLxxEExKxxxExQ	0.44	658—677	
RExHxxQxxxLRDExxxK	0.48	682—699	
LxDxxQxLxL	0.46	706—715	
DLKGLEETVxxELQTLHNLRKLFxxDLxxRxxK	0.53	749—781	No structure available, predicted C-terminal region.
KQKISFLENNxLEQLTKVHKQ	0.63	785—803	
LVRDNADLRCELPKLEKRLRAxxERVKALExALxEAKExAxxD	0.58	804—846	
YQxEVxRIKEAVRxK	0.52	849—860	

RxxxGxxAQIAKPxRxG		0.43	865—879

Table 13. Motifs detected in the kinesin-1 sub-family are listed together with their conservation, position and structure in the reference sequence *Neurospora crassa* kinesin-1

Outside the motor domain, 18 motifs have been detected in the coiled coil and the C-terminal region. However, these motifs are only detectable with hydrophobic/hydrophilic or polar/non-polar conservation criteria. It indicates that the coiled coil and the C-terminal regions have been strongly modified during evolution so that no sequence similarity can be detected any more. However, the observed changes presumably did not change the functional and biochemical properties of the motifs. Most positions in the motifs are over 80% conserved. A BlastP search of the motifs detected in the coiled coil and the C-terminal region was performed against the public database. All matches (e-value < 1) were kinesin homologues. This suggests that these motifs are kinesin-specific and might be important for kinesin structure or cargo binding. A search against the SMART [92,93] database has detected some matches with known protein domains (Table 14). However, the e-values are less significant than the required threshold of the prediction program. Further analysis of these regions could help to verify the predictions and determine the true properties ans functions of these motifs.

SMART match	Position	Evalue	Annotation
L27	526-598	4.28e+03	Domain in receptor targeting proteins Lin2 and Lin7
Microtub_assoc	533-621	2.8e+00	Proteins with this domain associate with the spindle body during cell division
BRLZ	561-614	1.33e+03	Basic region of leucin zipper
GIT	565-594	4.73e+02	Helical motif in the GIT family of ADP-ribosylation factor GTPase activating protein
Hr1	578-629	4.62e+03	Rho effectors or kinase-C related protein homology region 1

Table 14. Match results of the search of kinesin-motifs against the SMART database.

3.4.4 Phylogeny of kinesin-1

Kinesin-1 sequences are found in 86 eukaryotic species available in the database, covering every kingdom of the eukaryotes, including Protista, Fungi, Animalia and Plantae. Figure 41 shows the phylogenetic tree of the kinesin-1 family, which was obtained using Bayesian phylogenetic reconstruction.

It comprises four clear clades, each of which represents a single kingdom. The tree shows that there are fewer differences between fungal and animal kinesin-1 than between species of the chromalvoelata and plantae group. In other words, the kinesin-1 proteins in fungi and animals likely evolved from a more recent common ancestor. The ancestor of fungal and animal kinesin-1 should be a descendant of the ancestor of the planta, which evolved from the chromalvoelata group. The relationship shown in this tree is consistent with the generally accepted phylogenetic tree of life (Figure 40).

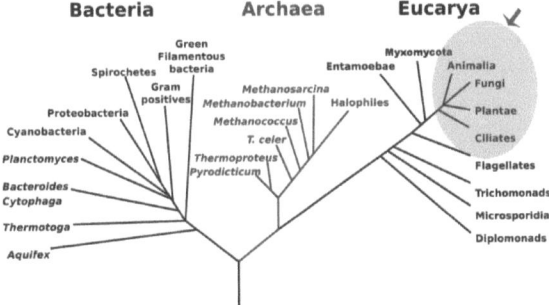

Figure 40. Phylogenetic tree of life based on rRNA data [94]. The region highlighted in green represents the relationship of the four kingdoms of eukaryotes.

Figure 41. Phylogenetic tree of the kinesin-1 sub-family with 129 sequences, obtained by mrbayes, 2 runs, 4 chains and 1 million generations. Every 100th tree was sampled. The first 1,000 trees were deleted as 'burn in'. This tree is the major consensus tree of 4,008 trees. The tree is rooted using the midpoint rooting method.

3.4.5 Fungal vs animal kinesin-1

During studies of the motility of kinesin-1 proteins, it has been observed that in the absence of cargo, kinesin-1 of several fungi moved about 3-4 times faster along the microtubule than animal kinesins. The phylogenetic tree indicates that kinesin-1 has split into several classes during evolution. Subsequently, these kinesins have been modified into group-specific proteins and they have become orthologs. The group-specific functions have been fixed in each group by a series of specific mutations, deletions or insertions during the evolutionary process. It is plausible that the difference in the motility of fungal and animal kinesin-1 is caused by some group-specific modifications of a common ancestor. During evolution, fungi required fast kinesin-1s for some reason, while the animals needed a slower version of kinesin-1. These requirements have been realized by specific modifications that became fixed within the members of each group. To recover these special modifications and eventually to find the trigger which controls the velocity of kinesin-1, it is necessary to do a comprehensive comparison of fungal and animal kinesin-1s.

Previous experimental evidence has suggested that the trigger is located inside the motor domain. Since a half-fungi/half-animal chimera of the motor domain did not change the animal slow kinesin-1 to a faster one, it is suggested that the trigger is more complex and presumably involves coordinated changes of several amino acids.

Closer inspection of the kinesin-1 phylogeny (Figure 42) shows that the sequences are grouped pretty much according to their taxonomic relationship. For example, in the animal group, insects and vertebrates fall into two clearly separated clades. In the vertebrate clade, three sub-clades can be observed. This strongly suggests a threefold duplication event within the vertebrate group. On the other hand, kinesin-1 sequences have been found in two of the fungal subkingdoms, namely the basidiomycota and ascomycota. The sequences belonging to the same class are clustered together as a sub-clade in the phylogeny. Intriguingly, the pezizomycotina, one subphylum of ascomycota, is more related to the basidiomycota than to the other two subphyla of ascomycota, the saccharomycetales and the *Schizosaccharomyces pombe*. They form a sister-clade to the pezizomycotina and basidiomycota. This association is supported by a significant posterior probability. It indicates that the tree is well supported and of high quality.

This tree was used as a guide tree for the prediction of ancestral sequences.

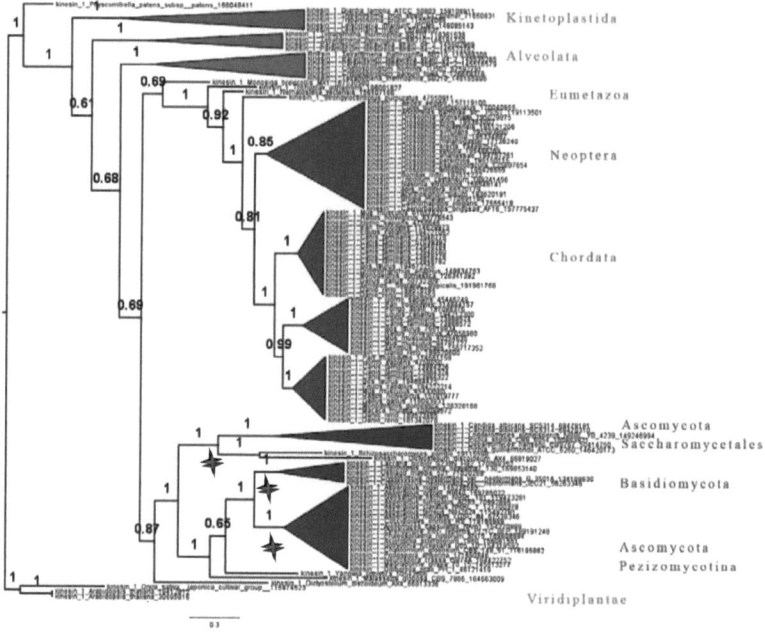

Figure 42. Phylogenetic tree of the kinesin-1 sub-family rooted at the plant branch. Sub-clades are colored. Fungal sequences and animal sequences are clearly separated into sister clades. Sequences within each clade belong to the same subphylum. Stars indicate where an ancestral kinesin-1 was synthesized.

3.4.6 Ancestral kinesin-1 prediction

Ancestral sequences were predicted for each interior node. Three of them, namely, the ancestor of fungal pezizomycotina, the ancestor of all animals and the ancestor of all fungi have attracted the most attention. If the biochemical and physical properties of these three ancestral proteins can be determined in the laboratory, important questions can be answered. For example, did the ancestral kinesin-1 have similar ATPase activity and microtubule binding ability? Did the ancestral kinesin-1 move along the microtubule similarly fast as the extant kinesin-1? Eventually, these ancestor sequences could help support the evolutionary tracing of sequence modifications.

The pairwise similarities of the motor domain of these ancestor sequences and two extant kinesin-1 proteins from the fungal and the animal group are shown in Table 15.

Two different evolutionary pathways are shown. One is Fungi_all → Fungi_p → NcK1; the other is Animal_all → MmK1. It shows that the ancestor of fungi is indeed the most distant sequence. Assuming the molecular clock is true, the following conclusions can be drawn. Firstly, the evolution of kinesin-1 within each group should be independent but convergent. That is, in both groups, the kinesin-1 sequences should be stabilized in structure and function during evolution. Because the sequence similarity S (Fungi_all, Fungi_p) = 94, S (Fungi_all, animal_all) = 86, but S (Fungi_p, animal_all) = 112 and S (Nck1, MmK1) = 107, it means that the sequences between groups are more similar to each other than to their common ancestor. Secondly, both the pezizomycotina group and the animal group are relatively young because the similarity S (Fungi_p, Nck1) = 166, S (animal_all, MmK1) = 145, while S (Fungi_all, Fungi_p) = 94 and S (Fungi_all, animal_all) = 86. Furthermore, the ancestor of animal kinesin-1 should be a little older than the ancestor of the pezizomycotina. Interestingly, the ancestor of fungi is a little bit more similar than the ancestor of pezizomycotina. It could be by chance that the NcK1 had changed several positions to the same amino acids as its ancestor, making them a bit more similar to each other than to the ancestor of pezizomycotina.

	Fungi_all	Fungi_p	Animal_all	NcK1	MmK1
Fungi_all		1e-94	2e-86	1e-97	2e-79
Fungi_p	1e-94		8e-112	7e-166	7e-111
Animal_all	2e-86	8e-112		5e-113	1e-145
NcK1	1e-96	7e-166	5e-113		2e-107
MmK1	2e-79	7e-111	1e-145	2e-107	

Table 15. Pairwise similarity of three ancestral kinesin-1s and two extant representatives of fungal and animal kinesin-1. Fungi_all: ancestor of fungi. Fungi_p: ancestor of fungal pezizomycotina. Animal_all: ancestor of animals. NcK1: kinesin-1 of *Neurospora crassa*. MmK1: kinesin-1 of *Mus musculus*. Similarity e-value is derived from the e-value of a blast2seq result.

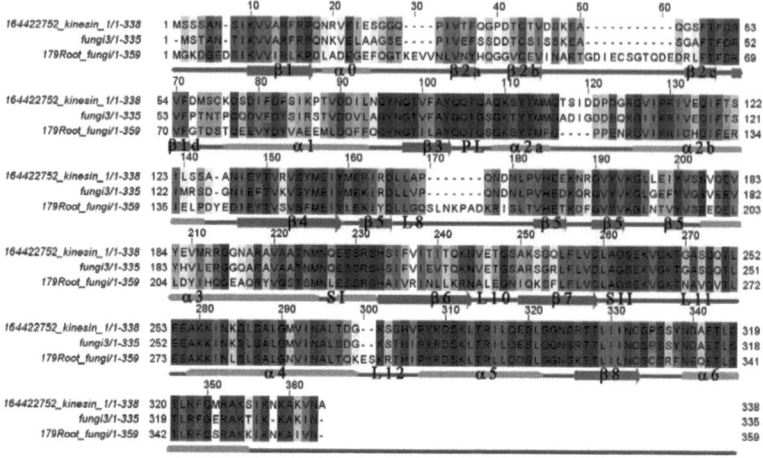

Figure 43. Alignment of motor domain sequences of the ancestor of fungi (179root_fungi), ancestor of pezizomycotina (fungi3), and kinesin-1 of *Neurospora crassa* (164422752_kinesin_1). Alignment is colored by the hydrophobicity of residues. The most hydrophobic residues are colored red and the most hydrophilic ones are colored blue. Secondary structure of 1GOJ is drawn under the alignment.

When comparing the ancestors of fungi with the kinesin-1 of *Neurospora crassa*, it is expected that the ancestors should have a similar secondary structure to extant kinesin-1. Indeed, most of the secondary structure elements are highly conserved. Five deletions have occurred during evolution from the ancestor of fungi to the ancestor of pezizomycotina. These changes were inherited by present-day kinesin-1s. One insertion appeared in the ancestor of pezizomycotina. This short insertion can be found also in the NcK1 sequence but with no conservation. All these indels are located in loop regions, making the corresponding loop shorter or longer. Known functional motifs, such as the p-loop, switch I, switch II, the ATP binding site and the microtubule binding site, are highly conserved.

The ancestor of animal kinesin-1 is highly similar to MmK1. Several small indels can be observed. As in the fungal sequences, these indels are all located outside essential regions (Figure 44).

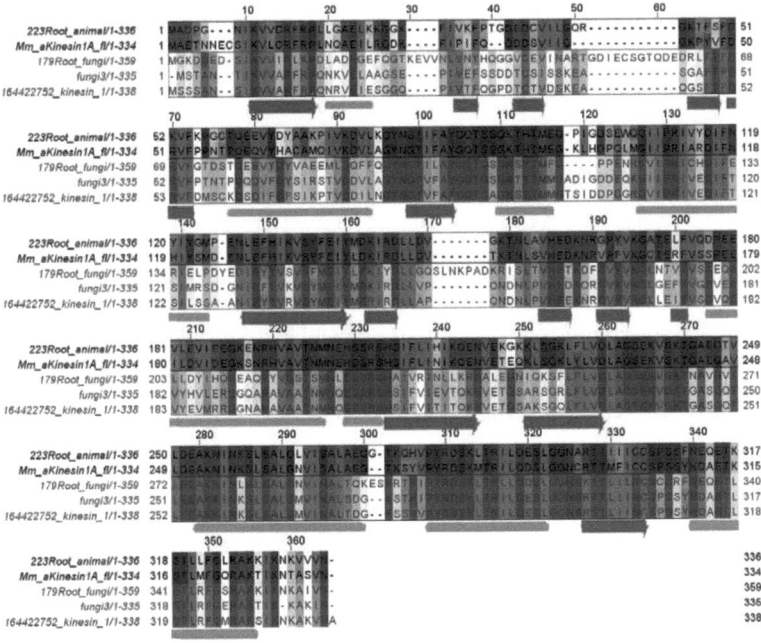

Figure 44. Alignment of animal and fungal ancestral kinesin-1 sequences.

On the other hand, there are many positions with differences in the alignment that show different degrees of conservation between groups. They are called discriminating positions and are thought to be important for the functional characteristics of each group, including the control of velocity.

The reconstruction of the ancestral kinesins performed from the bottom up in a step-by-step fashion. At present, four head domains of predicted ancestors have been synthesized. They include the ancestors of pezizomycotina, saccharomycetales / schizosaccharomyces, basidiomycota, and all animals. The three fungal ancestors have been tested in the laboratory by Marija Vukailovic (unpublished). All of them have been successfully expressed. They can hydrolyze ATP, bind to, and move towards the plus end of microtubules like extant kinesin-1 proteins. The motility tests indicate that the ancestors of fungal pezizomycotina and basidiomycota are both fast kinesins, their velocities are 1.9μm/sec and 1.7μm/sec, respectively. Interestingly, the ancestor of the saccharomycetales and schizosaccharomyces is a slower kinesin. Its speed was only 0.13µm/sec under same the conditions [95]. This is at odds with the hypothesis that all fungal kinesin-1s are fast and the ancestor of kinesin-1 is also a fast kinesin. The phylogeny of kinesin-1 indicates that this group is the oldest fungal group, probably evolving directly from the ancestor of the entire kinesin-1 sub-family. It might suggest that the ancestor of all kinesin-1s of fungi and animals is a slow kinesin. To test this idea, further experiments on the animal ancestor and the ancestor of all kinesin-1s are necessary.

3.4.7 Discriminating positions in fungal and animal kinesin-1

To investigate the sequence differences between fungal and animal kinesin-1, an alignment that only included fungal and animal kinesin-1 was created. 71 sequences from the animal group and 33 sequences from the fungal group were included.

The analysis of the alignment reveals 247 discriminating residues, which are over 80% conserved in the animal group, but differ from the fungal group (for the complete list, see supplementary material).

Among these discriminating positions, 22 are over 80% conserved in both groups (Figure 45). For example, position 91 of the NcK1 is 100% conserved in the fungal group with glycine. Its corresponding position 88 of HsK1 (the kinesin-1 of *Homo sapiens*) is 94% conserved in the animal group with serine. 8 residues are located in beta sheets (49β2c, 132β4, 134β4, 166-167β5c, 176β5d, 227β7, 306β8); 5 in helices (99α2a, 204α3b, 257α4, 270α4, 273α4) and 8 in loops (91p-loop, 144L7, 152L8, 154L8, 251L11, 253L11, 341neck, 345neck). One residue, 886R, is found in the C-terminus. In addition, the characteristics of positions 130, 140, 249, 336, and 340 remain unchanged between fungi and animals.

On the other hand, 23 residues are over 80% conserved only in the fungal group, but less than 80% conserved in the animal group (Figure 46). Among these, 5 residues are found in helices (63α1, 274α4, 287α5, 293α5, 322α6), 4 in sheets (30β2a, 139β4, 303β8, 304β8), and 10 residues in loops (143L7, 155L8, 172L8, 224L10, 250L11, 276L12, 277L12, 280L12, 299L13, 334neck linker). 4 residues (359, 472, 556, 893) are located outside the motor domain and are not shown in the list. One residue (P926) is located in the C-terminus. There are 9 amino acid residues whose biochemical characteristics did not change.

Animal			Fungi				
pos	grp	AA	con.	pos	grp	AA	con.	hydrophilicity	polarity
46	0	Y	0.90	49	3	F	0.85	neutral=>hydropho	polarUncharged=>nonpolar
88	0	S	0.94	91	3	G	1.00	neutral=>neutral	polarUncharged=>nonpolar
96	0	E	0.96	99	3	M	0.85	hydrophi=>hydropho	polarCharged=>nonpolar
128	0	F	0.99	132	3	Y	0.91	hydropho=>neutral	nonpolar=>polarUncharged
130	0	I	0.97	134	3	V	0.88	hydropho=>hydropho	nonpolar=>nonpolar
140	0	D	0.93	144	3	E	1.00	hydrophi=>hydrophi	polarCharged=>polarCharged
148	0	V	0.96	152	3	P	0.82	hydropho=>neutral	nonpolar=>nonpolar
150	0	K	0.99	154	3	N	0.85	hydrophi=>hydrophi	polarCharged=>polarUncharged
162	0	V	0.97	166	3	G	1.00	hydropho=>neutral	nonpolar=>nonpolar
163	0	P	0.99	167	3	V	0.82	neutral=>hydropho	nonpolar=>nonpolar
172	0	F	0.99	176	3	Y	0.82	hydropho=>neutral	nonpolar=>polarUncharged
200	0	H	0.99	204	3	E	0.91	neutral=>hydrophi	polarCharged=>polarCharged
223	0	L	0.96	227	3	K	0.94	hydropho=>hydrophi	nonpolar=>polarCharged
247	0	V	1.00	251	3	T	0.94	hydropho=>neutral	nonpolar=>polarUncharged
249	0	D	0.96	253	3	E	0.97	hydrophi=>hydrophi	polarCharged=>polarCharged
253	0	N	1.00	257	3	K	0.97	hydrophi=>hydrophi	polarUncharged=>polarCharged
266	0	S	0.96	270	3	N	0.94	neutral=>hydrophi	polarUncharged=>polarUncharged
269	0	A	0.97	273	3	T	1.00	neutral=>neutral	nonpolar=>polarUncharged
301	0	C	0.99	306	3	N	0.94	hydropho=>hydrophi	polarUncharged=>polarUncharged
336	0	T	0.94	341	3	S	1.00	neutral=>neutral	polarUncharged=>polarUncharged
340	0	W	0.97	345	3	L	0.91	hydropho=>hydropho	nonpolar=>nonpolar
919	0	Q	0.89	886	3	R	0.82	hydrophi=>hydrophi	polarUncharged=>polarCharged

Figure 45. List of discriminating residues of both the fungal and animal groups that are over 80% conserved. Lines highlighted in green are residues without a change of biochemical properties. The sequence positions refer to HsK1, gi4758648. The reference sequence of the fungal group is NcK1, gi164422752.

Animal			Fungi				
pos	grp	AA	con.	pos	grp	AA	con.	hydrophilicity	polarity
31	0	P	0.41	30	3	V	0.82	neutral=>hydropho	nonpolar=>nonpolar
60	0	Q	0.62	63	3	D	0.82	hydrophi=>hydrophi	polarUncharged=>polarCharged
135	0	F	0.75	139	3	M	0.91	hydropho=>hydropho	nonpolar=>nonpolar
139	0	L	0.62	143	3	M	0.91	hydropho=>hydropho	nonpolar=>nonpolar
151	0	T	0.66	155	3	D	0.82	neutral=>hydrophi	polarUncharged=>polarCharged
168	0	C	0.62	172	3	L	0.91	hydropho=>hydropho	polarUncharged=>nonpolar
220	0	E	0.63	224	3	G	0.85	hydrophi=>neutral	polarCharged=>nonpolar
246	0	A	0.59	250	3	Q	0.91	neutral=>hydrophi	nonpolar=>polarUncharged
270	0	E	0.72	274	3	D	1.00	hydrophi=>hydrophi	polarCharged=>polarCharged
271	0	T	0.42	276	3	K	0.94	neutral=>hydrophi	polarCharged=>polarCharged
272	0	K	0.73	277	3	S	0.85	hydrophi=>neutral	polarCharged=>polarUncharged
275	0	V	0.68	280	3	I	0.82	hydropho=>hydropho	nonpolar=>nonpolar
282	0	M	0.63	287	3	L	1.00	hydropho=>hydropho	nonpolar=>nonpolar
288	0	D	0.61	293	3	E	0.94	hydrophi=>hydrophi	polarCharged=>polarCharged
294	0	C	0.61	299	3	S	0.97	hydropho=>neutral	polarUncharged=>polarUncharged
298	0	I	0.77	303	3	L	1.00	hydropho=>hydropho	nonpolar=>nonpolar
299	0	V	0.70	304	3	I	0.97	hydropho=>hydropho	nonpolar=>nonpolar
317	0	M	0.42	322	3	R	0.97	hydropho=>hydrophi	nonpolar=>polarCharged
329	0	V	0.76	334	3	A	0.94	hydropho=>neutral	nonpolar=>nonpolar
357	0	T	0.39	362	3	Y	0.88	neutral=>neutral	polarUncharged=>polarUncharged
467	0	Y	0.51	472	3	L	0.82	neutral=>hydropho	polarUncharged=>nonpolar
551	0	I	0.46	556	3	K	0.82	hydropho=>hydrophi	nonpolar=>polarCharged
926	0	P	0.59	893	3	G	0.85	neutral=>neutral	nonpolar=>nonpolar

Figure 46. Discriminating residues that are over 80% conserved in the fungal group but less conserved in the animal group

The other 224 residues are over 80% conserved in the animal group, but less than 80% conserved in the fungal group. Among these, 55 residues are located in the motor domain. 21 of them are in helixes, 17 in beta sheets and the other 17 residues in loops

(Table 16). 17 positions without a biochemical property change are highlighted in Figure 47.

Structure element	Discriminating Residues
Alpha helix	(62,64,65,69,72,76)α1, (95,96) α2a,(118,124) α2b, (180,187,189,191,193,195,198) α3a, 267α4,(314,318,329) α6
Beta sheet	9β1,31β2a,47β2b,84β3,(130,133,135) β4,145β5a,162β5b,175β5d, (213,215,217,219) β6, (226,230,232) β7
Loop	17L1,92p-loop,126L6,151L8a,164L8b,173L8c, 203switchI, (243,248)switchII, (342,343,349,352,353,354,355)neck linker

Table 16. Location of the animal kinesin-1 – specific residues in different secondary structure elements.

These residues are almost equally distributed in the three major different structural elements. In helices, they are located mainly in α1, α3a and α6. There is no specially conserved animal residue found in α5, indicating that helix α5 may not be an indicator for the functional divergence between groups.

Together with the residues described above, which are conserved in both groups, 28 residues in the β sheets are potentially animal-specific. Among these, the most significant appear to occur in β4, β6 and β7. These β-sheets are mostly buried inside the protein and function as highly conserved scaffolds for the kinesin motor domain. Changes that occurred in this region could be important for the structural integrity of the motor domain.

Overall, most of the discriminating residues are found conserved in the animal group, but less or not conserved in the fungal group. Notice that there are 71 animal sequences sampled for the analysis, over two times more than fungal sequences. Furthermore, the phylogeny of kinesin-1 shows that the animal and fungal groups evolved independently since the split from their common ancestor. The average conservation of animal group kinesins is about 67%, while the fungal group (without the oldest saccharomycetales and *Schizosaccharomyces pombe*) is only 56% conserved. This suggests that the fungal group evolved in general faster than the animal group.

Animal			Fungi			hydrophilicity	polarity
pos	grp	AA	con.	pos	grp	AA	con.		
6	0	K	0.87	5	3	S	0.30	hydrophi=>neutral	polarCharged=>polarUncharged
18	0	L	0.87	17	3	Q	0.52	hydropho=>hydrophi	nonpolar=>polarUncharged
28	0	K	0.82	27	3	E	0.33	hydrophi=>hydrophi	polarCharged=>polarCharged
44	0	K	0.89	47	3	G	0.55	hydrophi=>neutral	polarCharged=>nonpolar
59	0	E	0.92	62	3	T	0.18	hydrophi=>neutral	polarCharged=>polarUncharged
61	0	V	0.96	64	3	I	0.67	hydropho=>hydropho	nonpolar=>nonpolar
62	0	Y	0.97	65	3	F	0.79	neutral=>hydropho	polarUncharged=>nonpolar
66	0	A	0.97	69	3	I	0.76	neutral=>hydropho	nonpolar=>nonpolar
69	0	I	0.97	72	3	T	0.79	hydropho=>neutral	nonpolar=>polarUncharged
73	0	V	0.96	76	3	I	0.58	hydropho=>hydropho	nonpolar=>nonpolar
81	0	I	0.92	84	3	V	0.76	hydropho=>hydropho	nonpolar=>nonpolar
89	0	S	0.96	92	3	A	0.55	neutral=>neutral	polarUncharged=>nonpolar
92	0	T	0.96	95	3	S	0.70	neutral=>neutral	polarUncharged=>polarUncharged
93	0	H	0.93	96	3	Y	0.79	hydrophi=>neutral	polarCharged=>polarUncharged
114	0	D	0.97	118	3	Q	0.73	hydrophi=>hydrophi	polarCharged=>polarUncharged
120	0	Y	0.97	124	3	L	0.48	neutral=>hydropho	polarUncharged=>nonpolar
122	0	M	0.97	126	3	S	0.76	hydropho=>neutral	nonpolar=>polarUncharged
124	0	E	0.80	128	3	S	0.55	hydrophi=>neutral	polarCharged=>polarUncharged
126	0	L	0.97	130	3	I	0.73	hydropho=>hydropho	nonpolar=>nonpolar
129	0	H	0.97	133	3	T	0.73	neutral=>neutral	polarCharged=>polarCharged
131	0	K	0.97	135	3	R	0.55	hydrophi=>hydrophi	polarCharged=>polarCharged
141	0	K	0.96	145	3	R	0.70	hydrophi=>hydrophi	polarCharged=>polarCharged
147	0	D	0.97	151	3	V	0.24	hydrophi=>hydropho	polarCharged=>nonpolar
158	0	D	0.99	162	3	K	0.67	hydrophi=>hydrophi	polarCharged=>polarCharged
160	0	N	0.99	164	3	S	0.45	hydrophi=>neutral	polarUncharged=>polarUncharged
169	0	T	0.99	173	3	L	0.55	neutral=>hydropho	polarUncharged=>nonpolar
171	0	R	0.99	175	3	V	0.45	hydrophi=>hydropho	polarCharged=>nonpolar
176	0	P	1.00	180	3	V	0.48	neutral=>hydropho	nonpolar=>nonpolar
183	0	I	0.99	187	3	M	0.61	hydropho=>hydropho	nonpolar=>nonpolar
185	0	E	0.96	189	3	R	0.55	hydrophi=>hydrophi	polarCharged=>polarCharged
187	0	K	0.99	191	3	G	0.64	hydrophi=>neutral	polarCharged=>nonpolar
189	0	N	0.99	193	3	A	0.67	hydrophi=>neutral	polarUncharged=>nonpolar
191	0	M	0.96	195	3	A	0.67	neutral=>neutral	polarUncharged=>nonpolar
194	0	V	0.99	198	3	A	0.55	hydropho=>neutral	nonpolar=>nonpolar
199	0	E	0.97	203	3	Q	0.48	hydrophi=>hydrophi	polarCharged=>polarUncharged
209	0	L	0.99	213	3	V	0.64	hydropho=>hydropho	nonpolar=>nonpolar
211	0	N	0.89	215	3	T	0.64	hydrophi=>neutral	polarUncharged=>polarUncharged
213	0	K	0.97	217	3	T	0.52	hydrophi=>neutral	polarCharged=>polarUncharged
215	0	E	0.97	219	3	K	0.76	hydrophi=>hydrophi	polarCharged=>polarCharged
222	0	K	0.93	226	3	A	0.52	hydrophi=>neutral	polarCharged=>nonpolar
226	0	K	0.99	230	3	Q	0.58	hydrophi=>hydrophi	polarCharged=>polarUncharged
228	0	Y	0.97	232	3	F	0.76	neutral=>hydropho	polarUncharged=>nonpolar
239	0	S	1.00	243	3	G	0.76	neutral=>hydropho	polarUncharged=>nonpolar
244	0	E	0.96	248	3	S	0.67	hydrophi=>neutral	polarCharged=>polarUncharged
263	0	N	0.96	267	3	M	0.73	hydrophi=>hydropho	polarUncharged=>nonpolar
309	0	E	0.80	314	3	D	0.48	hydrophi=>hydrophi	polarCharged=>polarCharged
313	0	K	0.96	318	3	L	0.61	hydrophi=>hydropho	polarCharged=>nonpolar
324	0	T	0.94	329	3	S	0.48	neutral=>neutral	polarUncharged=>polarUncharged
337	0	K	0.97	342	3	P	0.76	hydrophi=>neutral	polarCharged=>nonpolar
338	0	E	0.96	343	3	A	0.67	hydrophi=>neutral	polarCharged=>nonpolar
344	0	Y	0.94	349	3	L	0.76	neutral=>hydropho	polarUncharged=>nonpolar
345	0	E	0.94	350	3	K	0.27	hydrophi=>hydrophi	polarCharged=>polarCharged
347	0	E	0.94	352	3	A	0.64	hydrophi=>neutral	polarCharged=>nonpolar
348	0	K	0.92	353	3	Q	0.58	hydrophi=>neutral	polarCharged=>polarUncharged
349	0	E	0.92	354	3	S	0.18	hydrophi=>neutral	polarCharged=>polarUncharged
350	0	K	0.94	355	3	Q	0.55	hydrophi=>hydrophi	polarCharged=>polarUncharged

Figure 47. Discriminating residues that are over 80% conserved in animal kinesin-1 but less conserved in fungi

3.4.8 Discriminating residues and the motor domain structure

The discriminating residues are potentially important for the specific structure and function of the motors in animals and fungi. To uncover the potential functional roles of these residues, it is important and necessary to relate the sequence information to the structure of the motor domain.

There are 77 crystal structures of kinesins available in the PDB database. 5 of them are kinesin-1 motor domain structures. 1GOJ [96] is the motor domain structure of the *Neurospora crassa* kinesin-1, 2KIN [97] and 3KIN [98] are structures of *Rattus norvegicus*. And 1BG2 [99], 1MKJ [100] and 2P4N [101] are from *Homo sapiens*.

1GOJ (*Neurospora*) and 1BG2 (*Homo*) were used for the structural analyses described here. This is because their structures are relatively complete and present structural differences from the same state in the hydrolysis cycle (ADP-bound). Therefore, they are perfect candidates for analyzing the potential significance of the differences between fungal and animal kinesin-1.

3.4.8.1 The bond between 204E and 257K in NcK1

The structure comparison of 1BG2 and 1GOJ shows that both structures are highly conserved, except for two major structural changes in switch I and loop-11. These two regions are believed to undergo conformational changes during the hydrolysis of ATP [87-91]. The α3 helix of NcK1 has one additional turn compared to HsK1. At the other end of switch I, NcK1 has one helical turn less than HsK1. Loop-11 of NcK1 has two additional helical turns at the junction of loop-11 and the α4 helix [89]. These changes most likely occurred as a result of different conformational requirements.

In these two structures, there are two important residues: 204E and 257K in NcK1, which correspond to 200H and 253N in HsK1. These two residues are highly conserved discriminating residues (Figure 45). In NcK1, they are form a bond that locks switch I and loop-11 in a closed conformation. In HsK1, the binding of these two residues is broken and the helical turns of switch I and loop-11 are released. The motor domain is in an open conformation (Figure 48).

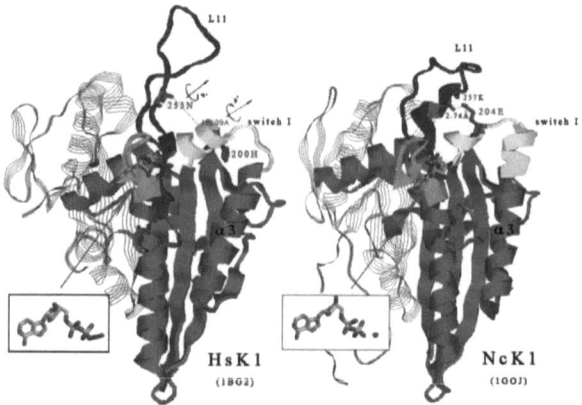

Figure 48. Structural comparison of animal kinesin-1 (HsK1), PDB ID: 1BG2, and fungal kinesin-1 (NcK1), PDB ID: 1GOJ. Both structures are highly conserved except for two major structural changes in the yellow-colored switch I and the blue-colored loop-11. In contrast, the green-colored p-loop (main component of the ATP binding pocket) is structurally equivalent. Helix α3 of NcK1 has one additional

turn compared to HsK1. At the other end of switch I, NcK1 has one helical turn less than HsK1. Loop-11 of NcK1 has two helical turns at the junction of loop-11 and the α4 helix not seen in HsK1. The two residues shown in red are highly conserved discriminating residues. In NcK1, they form a bond and lock switch I and loop-11 in a closed conformation. In HsK1, the binding of these two residues is broken and the helical turns of switch I and loop-11 are released. The motor domain is in an open conformation. The two small frames show the expanded ADP and Mg++ molecules.

3.4.8.2 Residue 203Q in switch I of NcK1

In the switch I region, there are two discriminating positions, 203 and 204. Position 204 was described above, locking switch I and loop-11 together by binding to residue 257K. Interestingly, the adjacent amino acid 203 is a discriminating position, too. Structure analysis shows that in NcK1, 203Q can form a bond with 140E of β4, the distance between them being 2.65 Ångstroms. The amino acid at position 140 is 98% conserved throughout the kinesin-1 sub-family.

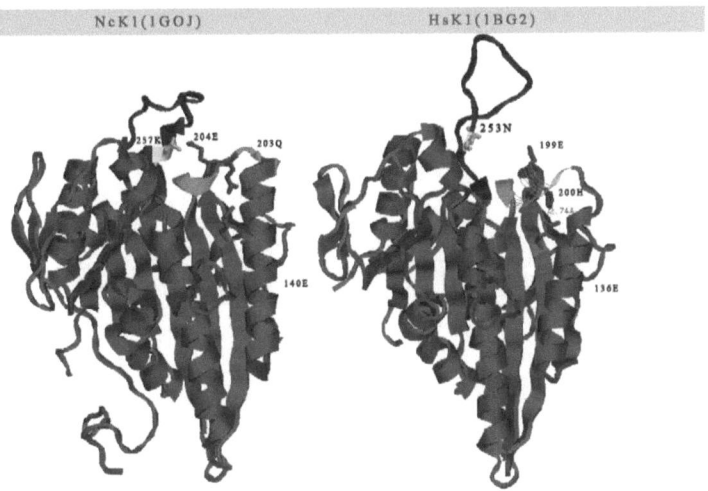

Figure 49. Structure comparison of a fungal (left) and an animal (right) motor domain. In NcK1, 204E (red) is bound to 257K (yellow) and 203Q (red) is bound to 140E (pink) in a closed conformation. In HsK1, the bond between 200H (red) and 253N (yellow) is open. 199E (red) is set free, and 200H (red) connects to 136E.

This binding could be important for stabilizing switch I. However, in HsK1, 136Q, which corresponds to fungal 140Q, cannot form a link to 199E (corresponding to fungal 203Q), but rather to 200E (fungal 204H). It seems that the position 204 acts as a two-way switch, while the position 203 as an assistant. When 204E binds to 257K to form the closed conformation form as described above, 203Q is used as an assistant to interact with 140Q, thus stabilizing switch I; otherwise, 204E binds 140Q in the open conformation state and 203Q is set free.

3.4.8.3 Residues 91, 92, 95, 96 in the ATP-binding pocket of NcK1

The ATP binding pocket is highly conserved in both structures. Most residues are over 95% conserved throughout the kinesin-1 family. However, there are four residues that fall into the discriminating position category. 91G (100% conserved in fungi) ⇔ 88S (94% in animals); 92A (55% conserved in fungi) ⇔ 89S (96% conserved in animals); 95S (70% conserved in fungi) ⇔ 92T (96% conserved in animals); and 96Y (79% conserved in fungi) ⇔ 93H (93% conserved in animals). For positon 93, 93% of the animal kinesin-1s have a basic histidine, while 79% of the fungal kinesin-1s have a tyrosine.

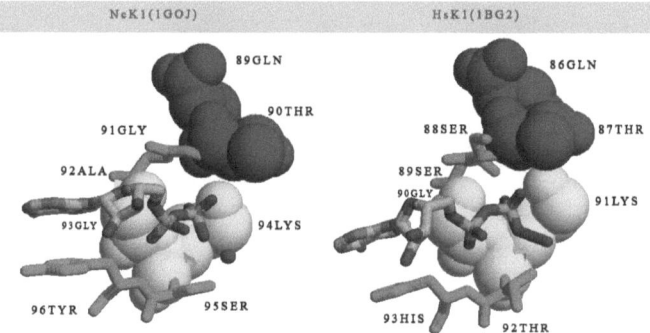

Figure 50. Comparison of the four discriminating residues between fungal and animal kinesin-1s. The figure shows the peptide 89-96 of NcK1 (1GOJ) and 86-93 of HsK1 (1BG2). Discriminating resides are colored in green. Red and yellow colored residues are conserved throughout the kinesin-1 family. The stick structure in the middle is ADP.

The tyrosine in fungi is only 79% conserved, it is only present in the pezizomycotina; the basidiomycota have phenylalanine at this position. These residues are similar in structure: a ring which forms a pi-interaction with the adenine ring of the ATP (Figure 50).

Phenylalanine and tyrosine are non-polar aromatic amino acids. It is known that tyrosine is derived from phenylalanine by hydroxylation in the para position. Hydrophobic tyrosine is significantly more soluble than phenylalanine, and its phenolic hydroxyl group is significantly acidic. When buried inside a protein, the ionization will yield an exceedingly unstable phenolate anion. Histidine has a positively charged imidazole group. It is used by many proteins in enzyme-catalyzed regulatory mechanisms, changing the conformation and behavior of the polypeptide in acidic regions [102]. It could probably be the reason why the histidine is fixed in the animal group, while the fungal kinesin-1s favor the tyrosine and phenylalanine.

The analysis of the velocities of fungal ancestor sequences synthesized in vitro showed that the ancestors of fungal basidiomycota and pezizomycotina are both fast. In these two groups, the sequence of the peptide containing the four discriminating positions (91, 92, 95, 96) in the ATP binding pocket is GSGKTF and GSGKTY, respectively. A comparison with NcK1, which has a GAGKSY motif, shows that the mutational combinations Y96F, (A91S, S95T) or (A91S, S95T, Y96F) do not lead to a slowing down of kinesin-1. This suggests that the velocity control is more complex and involves additional changes unrelated to the ATP binding pocket. In general, it is believed that a high ATPase activity can accelerate the ADP release step and thus lead to a high enzymatic velocity [96]. The changes in the four discriminating amino acids could work together to build a more stable ATP-binding pocket to slow down the ADP release and consequently affect the velocity of kinesin-1.

3.4.8.4 Residue 243 in NcK1

A previous study (unpublished; U. Majdic, PhD thesis 1999) has discovered a noteworthy difference in the switch II region between fungal and animal kinesin-1, where the glycine in fungal sequences is replaced by a serine in animal sequences (residue 243 in *Neurospora* kinesin-1). In a mutational analysis, a point mutation was introduced in NcK1 in this position, substituting the glycine with serine, the corresponding amino acid in animal kinesins. The gliding velocity of NcK1 with the animal SKT-motif was reduced to 73% of the wild-type velocity, whereas the motile behaviour of a *Drosophila* kinesin with the (fungal) GKT-motif was unchanged. Notice that residue 243 is located in a relatively open loop structure, where it has no lateral interaction with any other residue. It is unlikely that this position is important for the structural integrity of the motor core. Serine differs from glycine in a methyl and a hydroxyl group. Switching from glycine to serine can increase the hydrophilicity of the protein. However, it is not clear why the wild-type velocity is reduced in the fungal kinesin-1 when increasing the hydrophilicity at this position.

3.4.8.5 Microtubule binding sites

The PDB entry 2P4N is one complex structure together with the HsK1 (1BG2) and the microtubule α, β subunits. It is a nine angstrom cyro-EM map of nucleotide free state. The structure of the monomeric motor domain is well depicted, including a complete loop-11. It enables the study of the microtubule binding interface.

The structure analysis with a search with a distance limitation of 3.5 angstrom between the motor domain and the microtubule in the complex reveals 25 residues involving in potentially microtubule binding. These residues are from 6 different secondary structure elements of the kinesin-1 motor domain (Table 17).

Residue in HsK1	Location	Conserved value	Discriminating	Corresponding residues in NcK1	Binding sites in tubulins
E157	β5b	0.99		E161	
D158	β5b		X	E162(3)	M416:b
K159	L8	0.93		K163	
K237	switchII	0.98		K241	E413:a,414:b
V238	switchII	0.92		V242	
E244	Loop11		X	S248(3)	E113:a
G245	Loop11	1.00		G249	
A246	Loop11		X	Q250(2)	D116:a
V247	Loop11		X	T251(1)	D116:a,R156:a
L248	Loop11	1.00		L252	K112:a,I115:a
D249	Loop11		X	E253(1)	K112:a
E250	Loop11	0.99		E254	H107:a,Y108:a
A251	Loop11	1.00		A255	T109:a
K252	Loop11	0.99		K256	G412:a
N253	Loop11		X	K257(1)	G410:a,G411:a
I254	Loop11	1.00		I258	V409:a
S272	Loop12		X	S277(2)	Q434:b
T273	Loop12	no	no	S278	
Y274	Loop12	0.74		H279	R264:b,S430,E431
R278	α5	0.98		R283	H192:b,E196:b

K313	α6		X	L318(3)		E420:a
S314	α6	0.94		S319		
L317	α6		X	R322(2)		E415:a
F318	α6	0.98		F323		
R321	α6	0.99		R326		R402:a

Table 17. Microtubule binding sites. Corresponding positions in HsK1 (1BG2) and NcK1 (1GOJ), their locations in the motor domain and binding sites of the microtubule(shown are residues with a distance smaller than 3.0 angstroms) are listed. One position can be conserved or discriminating. When conserved, the absolute conservation value is shown. The detailed information of a discriminating position is shown in previous lists. The number in brackets refers to the corresponding discriminating position list: 1: Figure 45; 2 : Figure 46; 3 : Figure 47. For the binding sites of the microtubule, the letters 'a' and 'b' appended to the residues stand for alpha tubulin and beta tubulin, respectively.

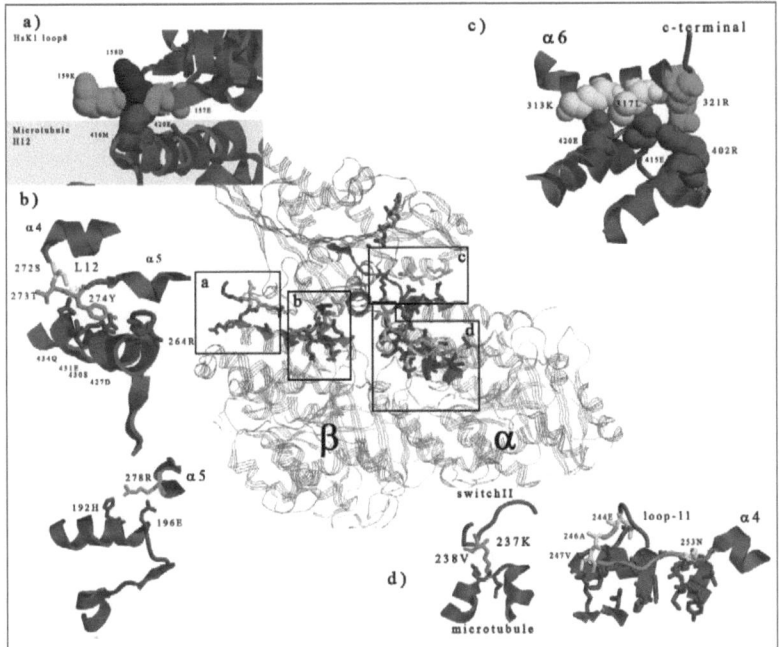

Figure 51. Potential microtubule binding sites obtained by structure analysis of 2P4N. The high resolution complex of a human kinesin (1BG2) docked on a microtubule is shown. Residues with a spatial distance smaller than 3 angstrom away from the microtubule interact with the microtubule. The search result indicates that the kinesin motor domain binds the microtubule in four microtubule regions. Two of them are located in β tubulin (a and b) and two in α tubulin (c and d). Connection targets in the

microtubule are depicted in red. Binding sites in the motor domain are colored in green, yellow and blue. Green colored residues are absolute conserved residues in the kinesin-1 family. Yellow and blue colored residues are discriminating residues (Table 17).

Nine residues are discriminating. The others are conserved in the kinesin-1 family with the exception of T273 in the loop-12. When reducing the search radius to 3.0 angstrom, the T273 is not included in the result any more. It means that the T273 is unlikely a part of the microtubule binding site.

All nine discriminating residues bind apparently directly to the microtubule, because the distance between one discriminating residue and their connection partner in the microtubule is less than 1.9 angstroms. It suggests that the discriminating residues could play a crucial role as binding regulators, while the absolute conserved residues are important for stabilizing the structure or creating necessary bindings with the microtubule.

Figure 51 shows that the potential binding sites of the microtubule can be grouped in four separate zones, like four legs sitting on the microtubule. Three of them, the β5b, loop-12-α5 and α6, are found on the rear side of the microtubule, which is the probable docking region of the neck linker (refer to the figure). The switch II and loop-11 are sitting at the front site. The binding of the loop-11 should be stronger than the others; because there are at least 10 residues involved in the binding to the microtubule with a distance less than 1.9 angstroms. In contrast, only one residue in β5b, one in α5, two in loop-12 and three in α6 are apparently connected with the microtubule.

Intriguingly, when comparing the binding sites in the loop-11 with those in the NcK1 structure, it demonstrates that the loop-11 in 1GOJ is in an inactive state. It is because these residues are structured significant differently in 1GOJ. First, the loop-11 is connected to the switch I (previous depicted), preventing the interaction of switch I and the microtubule. Second, the sites 248-252, the potential microtubule binding sites, are enclosed in the horseshoe structure by the connection between residue 244 and 252 (240 and 248 in HsK1) (Figure 38). This structure disables these resides for binding to the microtubule. Furthermore, the sites 255-258 in NcK1 (251-254 in HsK1) are structured as extended helical turns of the α4. They are unlikely to create a strong connection with the microtubule.

Notice that the highly conserved residue 240 (244 in NcK1 to close the horseshoe) does not appear in the binding sites. The connection search shows that no residue appears in its 3.5 angstroms radius area. It indicates that the only role of site 240 in the kinesin-1 motor domain is to close the horseshoe in closed conformation state. On the other hand, the other residue used for closing the horseshoe structure, the site 248 (252 in NcK1) is connected to the microtubule, showing its second distinct function.

3.4.8.6 Residues I130 and V173 in the HsK1

There are 5 Discriminating residues in β4 strand. The search for connected residues within 3.0 angstroms delivers one interesting candidate, the 130I in the HsK1 (V134 in the NcK1). It is connected with the V173 (V177 in the NcK1), a 90% conserved residue in the kinesin-1 family. In contrast, the corresponding pair V134 and V173 does not interact (Figure 52). When comparing the different conformation states of two motor domains, it shows that this pair of residues is one stabilization point for the α3. It could also be important for locating the β5-loop8 to the right binding position in the open conformation state.

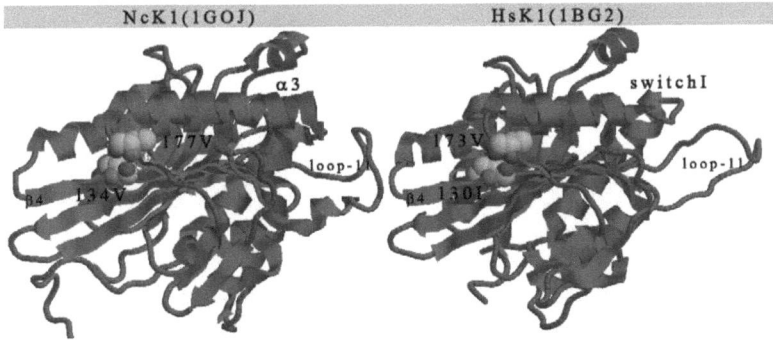

Figure 52. Pair of V134 and V177 (I130 and V173 in NcK1) unconnected in the closed conformation (NcK1), but connected in the opened conformation (HsK1), stabilizing the α3 helix.

3.4.8.7 Residues K226 and D288 in the HsK1

Another functional pair of discriminating residues is the K226 and D288 in the HsK1, corresponding to the Q230 and E293 in the NcK1. D288 is a discriminating residue located in the α5. K226 is found in a 3.0 angstroms radius zone of D288, together with the other residue N78 (N81 in NcK1) from the loop-4 region, which is a 95% conserved residue in the kinesin-1 family. The K226 is one residue of the β7 and the N78 is one residue of the loop-4. The structure of the HsK1 (1BG2) suggests that they are bonded. However, the N81 and Q230 do not interact with E293 in the NcK1 structure (Figure 53). This may be due to the conformational change of the motor domain between the open conformation (1BG2) and the closed conformation (1GOJ). When the motor domain is in the open state, the binding of the N78 and K226 with the D288 should be able to fix the α5 helix. It could be a necessary binding for the motor domain in order to help the R278 of the α5 helix to bind to the microtubule (Figure 51). Notice that the discriminating position 226 is over 99% conserved in the animal group and only 58% conserved in the fungal group, while the position 288 is 94% conserved in the fungal group and only 64% conserved in the animal group. The predicted ancestor sequence of the animal has also a lysine at this position, and the ancestor of fungi has a phenylalanine. It suggests that the K226 mutated in the animal group after the animal-fungi split. It must have some special function because of its 99% conservation (six different residues were observed at this position in fungi). One possible explanation could be that in fungi and their ancestor only the N81 is necessary for binding to the α5 helix in the open conformation, while the animal kinesin-1s need one additional residue to accomplish this task. The additional residue enhanced the binding with the α5 helix. It could be a necessary step for a more stable binding between the α5 helix and the microtubule.

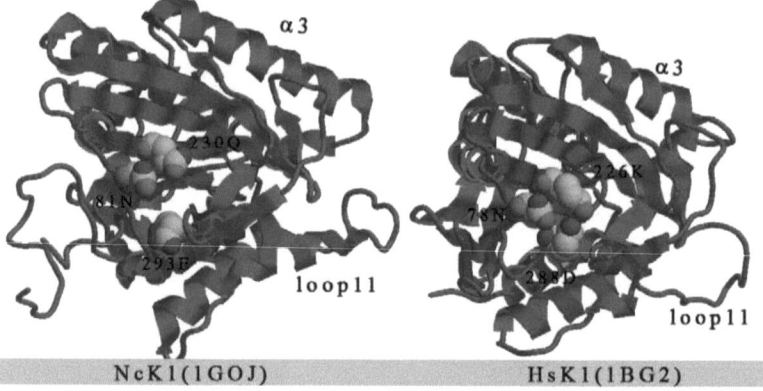

Figure 53. The discriminating pair of residues D288 and K226. Together with N78 they are connected in the open conformation state, linking the α5 helix and β7 and loop-4 together (right); they are disconnected in the closed conformation state (left).

3.4.8.8 Residues S266 and R321 in the HsK1

The S266 is one of three discriminating residues in the α4 helix in the HsK1, 96% conserved in the animal group. Its corresponding residue in the NcK1 is the T270, 94% conserved in the fungal group. The structure comparison shows that it binds the R321 of the α6 helix in the HsK1, but is disconnected with the R326 (corresponding residue in NcK1 of the R321) in the NcK1 (Figure 54). It indicates that the function of this residue is to fix the α4 and α6 in the open conformation state.

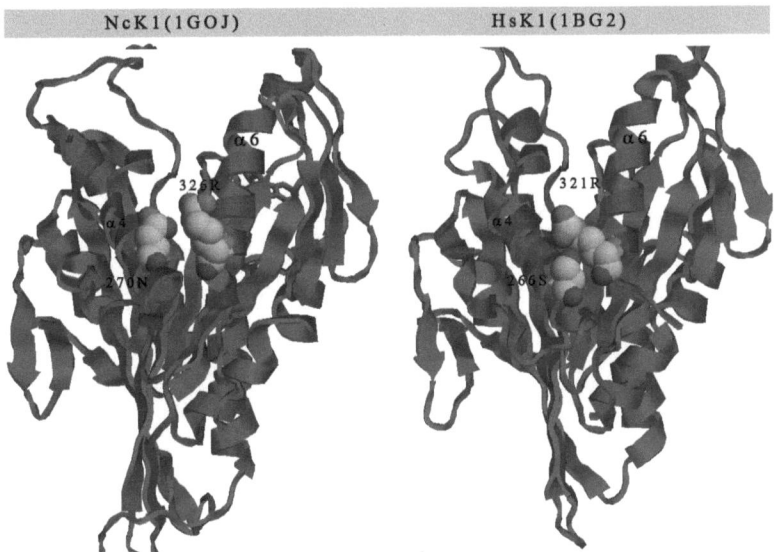

Figure 54. The residues S266 and R321 are connected in HsK1, the corresponding pair in the NcK1, N270 and R326, is not connected.

3.4.8.9 V148 and D144 in the HsK1

V148 is one discriminating residue found in the loop-8 of the HsK1. It is 96% conserved in the animal group and 82% conserved in the fungal group. Its corresponding residue in the NcK1 is P152. The structure analysis indicates that this residue is the key residue responsible for locking and releasing the loop-8-β5bc. In the closed conformational state of the motor domain, it connects the 98% conserved residue D144 (D148 in the NcK1) from the β5a to lock the loop-8a-β5b-loop8b-β5c region, which is in turn fixed to the β-sheet core of the kinesin motor domain through continuous bindings: D144 (β5a) ➔T169 (loop8c)➔S133 (β4)➔L209 (β6). In the open conformational state, the V148 and D144 are disconnected, so that the loop-8a-β5b-loop8b-β5c is set free in order to bind the microtubule (It could be also possible that the binding with the microtubule breaks the D144-V148 binding). The bindings D144 (β5a) ➔ T169 (loop8c) ➔ S133 (β4) ➔ L209 (β6) remain unchanged, so that the rest part of the loop8-β5 is still fixed on the motor domain. Furthermore, the fixation is enhanced by an additional binding between the I130 and V173 (previous depicted Figure 52)

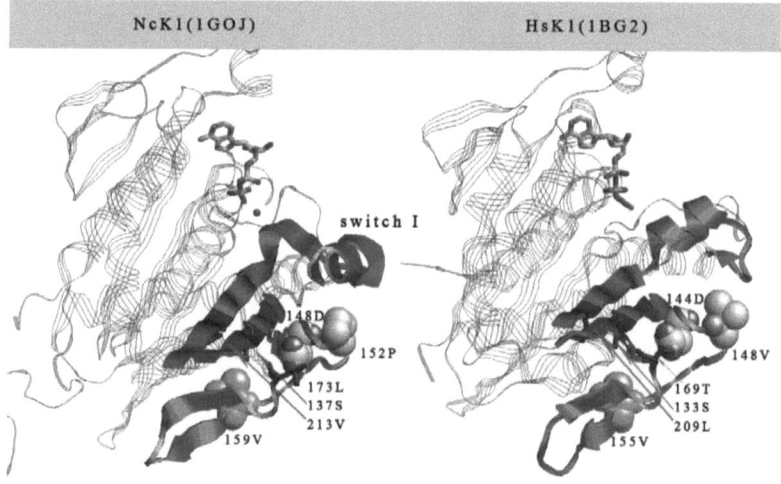

Figure 55. P152 and D148 are connected in the closed conformation state of the NcK1 (1GOJ). The corresponding pair of residues in the HsK1, the V148 and D144, are disconnected in the opened conformation state, so that the region loop-8a-β5b-loop8b-β5c, which contains the microtubule binding site V155 (green), is set free. The remaing part of loop8-β5 is fixed on the β core of the motor domain, independent of the conformational state. The fixation is generated by the blue colored residues. D144 (β5a) ➔T169 (loop8c)➔S133 (β4)➔L209 (β6).

3.4.9 Putative cooperative residues combinations

The chimera experiments (see introduction) have shown that the velocity of the animal kinesin-1 motor domain did not change by substituting its second half (including the microtubule binding interface, switch I and II) of the fungal kinesin-1 motor domain. However, the reverse chimera (fungal first half+ animal second half) did reduce the motor's velocity. The first experiment indicates that the velocity is not controlled by the second half of the motor. It should behave similar like the animal's second half. However, the latter one demonstrates that the first half of the motor is not able to control the velocity alone. The reduced speed of the chimera motor suggests that substituting of the fungal first half could break some important structure bindings, so that the motor domain cannot efficiently change the conformation like the wild type and does not become faster as expected, but even slower.

It demonstrates that there should be some important interaction between the first half and the second half of the motor domain, making the structure compact and function most efficient.

A search for such possible interactions was done for both NcK1 and HsK1 motor domains. The results are shown in Table 18. The putative connection pairs are grouped into three different classes: NcK1 only, HsK1 only or both together.

Eleven connections are found only in the NcK1. Six of them are located in the motor domain. The other 5 are found in the neck linker region; however, this part of the structure is missing in the HsK1 structure so that no comparison could be done for these five residues. The six connected pairs in the NcK1 are found in the HsK1 disconnected. For example, the corresponding pairs of pair 184-117 and pair 188-104 in the HsK1 are disconnected. It suggests that these two pairs are the linkages of the α3a helix and α2 helix (loop5 locates inside of α2) in the closed conformation. Another interesting pair is the 236-94, because the 236 is located in the switch II and the 94 is in the ATP binding pocket. In the closed conformation, the distance between them is 2.9 angstroms, while in the open conformation, the distance becomes 3.14 angstroms. It indicates that the structure of the motor domain becomes looser when the conformation changes from the closed state to the open state. Similar change can be observed for the connection pair 84-232-82. It is a β sheet bond between β3 and β7, the distance between them is changed from 2.8 angstroms to 3.1 angstroms.

Five connections are found only in the HsK1, in the open conformation state. The connection pair 78-288-226 has already been mentioned above (Figure 53). Intriguingly, the pair 224-76 is found connected in the open conformation state. They are adjacent to the connected pair 227-78 in the NcK1. The distance between them in the closed conformation is 3.28 angstroms and in the opened state is 2.9 angstroms. When the conformation changes to the open state, the four putative connected pairs between the β3 and β7 in the closed conformation state are disconnected and the connection between the pair 224-76 is formed. It should be used for binding the β3 and β7 together in the open conformation state.

The other three connections are found in the loop0 region. The N-terminus of the protein is connected with the α4 and loop13. Notice that the neck linker region in the closed conformational state is bond with the N-terminus. Although there is no structure information available for the neck linker region in the open conformation, it can be speculated that at least the neck linker is not bound to the motor domain any more in the open state. In fact, while the other regions of the motor domain bind the microtubule in the open state, it is likely that the neck linker is docked on the microtubule as well.

In addition, there are 15 connections found in both structures. These connections are conformation independent and are used as necessary linkages for stabilizing the native structure of the motor domain. For example, it shows that the β3 and β7 are linked through three conserved connections and the β8 is linked to β1 and β3 through three conserved connections as well. Intriguingly, some residues bind different number of residues when the conformational changes. For example, the β7 interacts with α5 in the open conformation (pair 226-288). The switch II binds two residues of the p-loop in the open conformation but only one in the closed conformation. The α6 is connected with the α0 and loop-14 in the closed conformation, but not in the open conformation. It suggests that the α6 is held by the α0 and loop-14 in the closed conformation but is set free to bind the microtubule in the open conformation.

NcK1(1GOJ)			HsK1(1BG2)
Connection found only in the NcK1 but not in the HsK1			
Res.Nr	Location	within 3.0 Å	
184	α3a	117	α2a
188	α3a	104	loop5
227	β7	78	loop4

		339	neck link				
232		82	β3				
		84	β3				
236	switchII	94	α2a				
314	α6	42	loop3				
		44	loop3				
331	Neck	4	loop0				
332	linker	6	loop0				
333		6	loop0				
334		3	loop0				
338		78	loop4				

Connection found only in the HsK1 but not in the NcK1

Res.Nr	Location	within 3.0 Å	
224	β7	76	loop4
268	α4	6	loop0
288	α5	78	loop4
		226	β7
292	Loop13	3	loop0
293		3	loop0

Common connections found in both structures

218	β6	77	end of α1	214	β6	74	
		227	β7			223	
230	β7	81	end of loop4	226	β7	78	loop4
		82	begin of β3			79	β3
						288	α5
234	β7	84	β3	230	β7	81	β3
		86	β3				
235	β7	95	α2a	231	β7	92	α2a
		210	β6			206	β6
240	switchII	90	p-loop	236	switchII	87	p-loop
						86	p-loop
301	β8	9 β1,294 α5		296	β8	8 β1,10β1,80β3	
302		83 β3,85 β3		297		82β3	
303		9 β1,11 β1		298		10β1,12β1	
304		85 β3,87 β3		299		82β3,84β3	
305		11 β1		300		12β1,14β1	
306		87 β3		301		84β3	
307		13 β1		302		14β1	
308	Loop14	21 α0		303	Loop14	22 α0	
		317 α6					
310	Loop14	17 end of loop1		305	Loop14	27 loop2	
311	Loop14	27 loop2		306	Loop14	28 loop2	
313	α6	21 α0		308	α6		

89 p-loop	86 p-loop

Table 18. Putative connected residues between the first half and the second half of the motor domain. They are grouped into 3 classes: only in NcK1 connected; only in HsK1 connected and in both connected.

3.4.9.1 The second layer of the ATP-binding pocket

The connections involved in the switch II and the ATP-binding p-loop attract most of the attention. The search for connected residues with the ATP-binding pocket, which locates in the first half of the motor domain, shows that the pocket is stabilized by several surrounding residues. In the NcK1, the pocket is surrounded by six residues from the other secondary elements (five connected residues are adjacent residues to the pocket and are not labeled in the figure), while in the HsK1 only five of them were detected. The connection L236-K94 in the NcK1 is disconnected in the HsK1 (the L232-K91) (Figure 56).

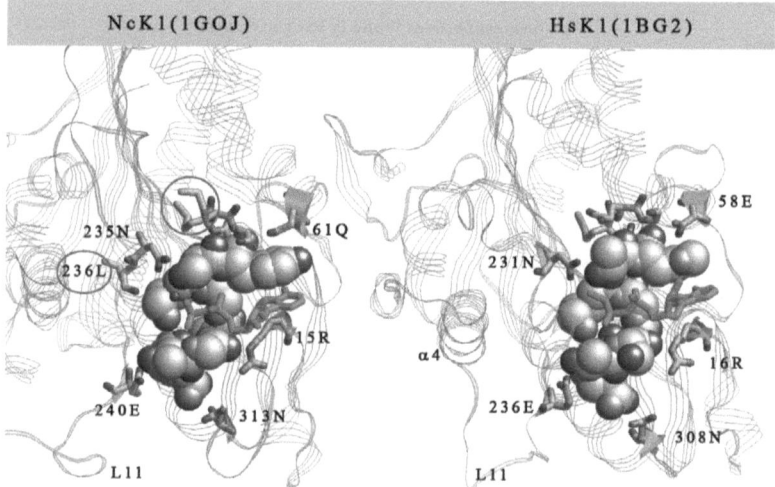

Figure 56. The ATP binding pocket of both NcK1 and HsK1 is surrounded by several highly conserved residues. All of these residues have a distance <3.0 angstroms to the binding pocket. The pocket of HsK1 is surrounded by five residues, while the one of NcK1 is surrounded by 6 residues. The 236L is disconnected with the binding pocket in the opened conformation state. Two methionines were observed in the loop5 in fungi, while there is only one present in animals. The additional methionine is a discriminating residue for fungi.

Of the six connections, two of them are from the first half, and the other four from the second half of the motor domain. All of them are highly conserved in the kinesin-1 family. Moreover, two residues of the p-loop, the A92 and G93 (S89 and G90 in the HsK1) have no contact with any residues from other structure elements except the p-loop itself. The second layer structure of the ATP-binding pocket suggests the internal compact interactions of entire motor domain structure.

In addition, there are apparently four residues in the neighboring loop5 connected to the pocket. In the NcK1, there are two methionines present, while in the HsK1 there is only one. The one present in both structures is 97% conserved. The other one is a discriminating residue, which is 90% conserved in the fungal group as methionine (M) and 96% conserved in the animal group as glutamic acid (E). It changed from a hydrophobic and non-polar (M) to hydrophilic and polar-charged (E) amino acid. The side chain of methionine is hydrophobic and the sulfur can react with electrophilic centers. It is known that the side chain of methionine is unbranched, providing considerable conformational flexibility [103]. The fungal kinesin-1 has two conserved methionines in the loop5, which could provide a larger non-polar surface and could be important for enzymatic functions.

3.4.10 Conjecture about the velocity controller

The p-loop is located in the first half of the kinesin-1 motor domain (1-124 of the HsK1). It is the main component of the ATP-binding pocket. The second half of the motor domain (125-325) contains the switch I and switch II motifs. The analysis of the microtubule binding sites indicates that all four putative microtubule binding regions locate in the second half of the structure (Table 17). This suggests that the first half is responsible for the ATP activity, while the second half is relevant for the microtubule binding and conformational changes.

3.4.10.1   ATP-binding pocket could be one candidate

It is believed that the ADP releasing speed determines the motor velocity. Thus, the ATP-binding pocket should be the first candidate to be considered for the speed control. To change the animal ATP-binding pocket to fungal one, the four discriminating residues should be relevant. The analysis showed that although all four residues are 96% conserved in the animal group, only the position S88 is also highly conserved in the fungal group. For the position S89, 45% of the fungal sequences have the same residue, the serine, at the corresponding position. For the position T92, 30% of the fungal sequences also have threonine at the corresponding position. And for the position H93, 79% of fungal sequences contain tyrosine and the others phenylalanine. Thus, the S88 and the H93 should be the most relevant residues responsible for the difference of the binding pocket. Mutating these two residues can easily switch the animal ATP-binding pocket to the fungal one.

The analysis of the ATP binding pocket shows that the pocket is actually supported by a second layer comprising six residues (Figure 56). These residues could be important for stabilizing the pocket structure (discussed above). Because these residues are all highly conserved in the kinesin-1 sub-family, substituting one or few residues of the ATP binding pocket should have no consequence for the binding pocket structure.

Furthermore, one discriminating residue is found in the loop5 at position 99 in the NcK1 (96 in the HsK1). This position is hydrophobic and non-polar in the fungal kinesin-1s (mainly methionine and leucine), but is hydrophilic and polar charged in the animal kinesin-1s. It could be important for the ATP binding or have an enzymatic function. Thus, this residue should be considered in order to obtain the full functional ATP binding pocket when doing mutagenesis.

3.4.10.2   Other potential factors for the velocity difference

The protein chimera experiments suggest that the second half of the motor domain alone is unlikely to be responsible for the different velocity between fungi and animals, because the chimera constructed from the first half of animal and the second half of fungi had similar speed like the wild type animal kinesin-1. On the other hand, the speed of the chimera's motor was reduced significantly by exchange of the first half of the animal motor domain with the fungal one.

One possible explanation could be that the supporting bindings with the second layer could be destroyed, although the main structure of the fungal ATP binding pocket remains unchanged. In this case, the binding pocket could become unstable without the full supporting of the second layer, causing an inefficient ATP binding and hydrolysis and affecting the velocity. If this is not the case, assuming that the structure of the ATP binding pocket did not change, the reducing of the speed indicates that the second half of the animal motor domain may control the speed because it can make a fast kinesin slow. The site-mutation experiment of G243S in the NcK1 motor domain has supported this hypothesis, in which the speed of the fungal mutant reduced to 73% of the wild type. However, introducing the entire fungal second half into the animal motor domain could not make the animal kinesin-1 faster. It suggests that the fungal ATP-binding pocket should be also necessary for controlling the speed.

In conclusion, the velocity controller should be a complicated cooperation of the ATP-binding pocket and the other important motifs. The fungal version of the ATP binding pocket is necessary for making the kinesin fast. Moreover, many special mutations in other functional or structural motifs may be effective too, such as the G243S in the NcK1.

Potential amino acid positions with goroup-specific functions (discriminating residues) are listed in Figure 45, Figure 46 and Figure 47. Figure 45 contains discriminating residues important for both groups. Figure 46 and Figure 47 contain fungal specific residues and animal specific residues respectively. These residues normally work in cooperation with other residues, such as the 204-257 connect switch I and loop-11 together.

Furthermore, several important pairs have been detected. They play important roles during the conformation change. It should possible to guide mutagenesis experiments in the future to test various combinations involved in velocity control.

106

4 Conclusion and discussion

4.1 Confidence evaluation of the phylogenetic tree

In this project, the dataset of kinesin sequences was expanded to over 2900. 2530 kinesin sequences were used for large phylogenetic tree estimation. For this large dataset, phylogeny inference is really a computational challenge.

Neighbor joining (NJ) [104] was used for similar work in other studies. For example, the myosin family tree with about 2,200 sequences was generated by this method [105]. However, the quality of tree inferred by this method is strongly dependent on the dataset, because the algorithm is based on the evolutionary distances between sequences. The distance between sequences is often related to the observed changes within the sequences. However, the observed mismatches between sequences do not always equal to their evolutionary distance. For example, multiple mutations could occur at the same site and make the sequences appear 'similar' to each other. In the case of kinesins, the motor domains are highly conserved, but still divergent to each other. The chance that two sequences are improperly clustered together is relatively high, especially for a large dataset. Therefore, the NJ method is unable to infer a correct phylogeny for the entire kinesin super-family.

In contrast, the maximum likelihood method and the alternative Bayesian inference method provide more realistic models for discovering the evolutionary relationships between sequences that have been separated for a long time. However, these methods are very time consuming and require a lot of computational resources. In practice, it has not been possible to use Bayesian methods such as mrbayes, to generating such a large phylogeny.

In this word, a phylogenetic tree has been successfully constructed by using RaxML, a maximum likelihood method implementation. However, no confidence testing like bootstrapping was applied to the tree. In this case, a confidence test of the tree was done by comparing the tree clades with the profile classification results.

To do so, the sequences were classified using position specific profile matrixes. The classification results were mapped on the tree, so that it is possible to calculate a relative quality score for each clade. The likelihood of a clade can be described by the proportion of sequences in that are from the same class. (See result).

This confidence measurement indicates that most clades of the tree were correctly estimated and the overall accuracy of the phylogenetic tree is about 85.7%.

4.2 The 14th kinesin sub-family

The standard nomenclature claims that there are 14 sub-families of kinesin. Another study has even found 17 different classes of kinesin based on a phylogenetic tree of about 400 sequences [17].

13 families are clearly supported by the large phylogenetic tree. Besides that, there are several small clades present in the tree. For example, the clades kinesin1.1, kinesin4.1 and kinesin7.1 are 100% supported, but do not fall into the same clade of their main classes. Intriguingly, all of the 31 sequences of the kinesin1.1 clade belong to the oldest eukaryotes, such as the kinetoplastids, green plants and *giardia lamblia* etc. The relationships among these sequences are consistent with the standard taxonomic tree. The 37 sequences of the kinesin4.1 clade are from the kinetoplastids and fungi, while the main clade of kinesin4 does not contain these two groups. It is the same case for the kinesin7. The clade kinesin7.1 is a pure fungal branch, which is lacking in the main clade of the kinesin7. Several phylogenetic attempts with other algorithms were applied on these parts of the dataset, but the relationships of these clades remain robust among all trees. It could indicate the splitting events in the early stage of these kinesin sub-families.

The other small clades are formed mainly by the sequences of kinesin-12, but also many members of other groups. When the 14th family of kinesin exists, it must be a part of the kinesin12 and forms a single clade with good support value in the first place. Additionally, the members of that clade should able to represent the distribution of the kinesin family among species.

The best supported clade is adjacent to the kinesin1. It is comprised of 92 sequences, 67 (73%) of them are from the kinesin-12 sub-family. The members are relatively widespread in the kinetoplastids, plants, and metazoa. No fungal species is found in this

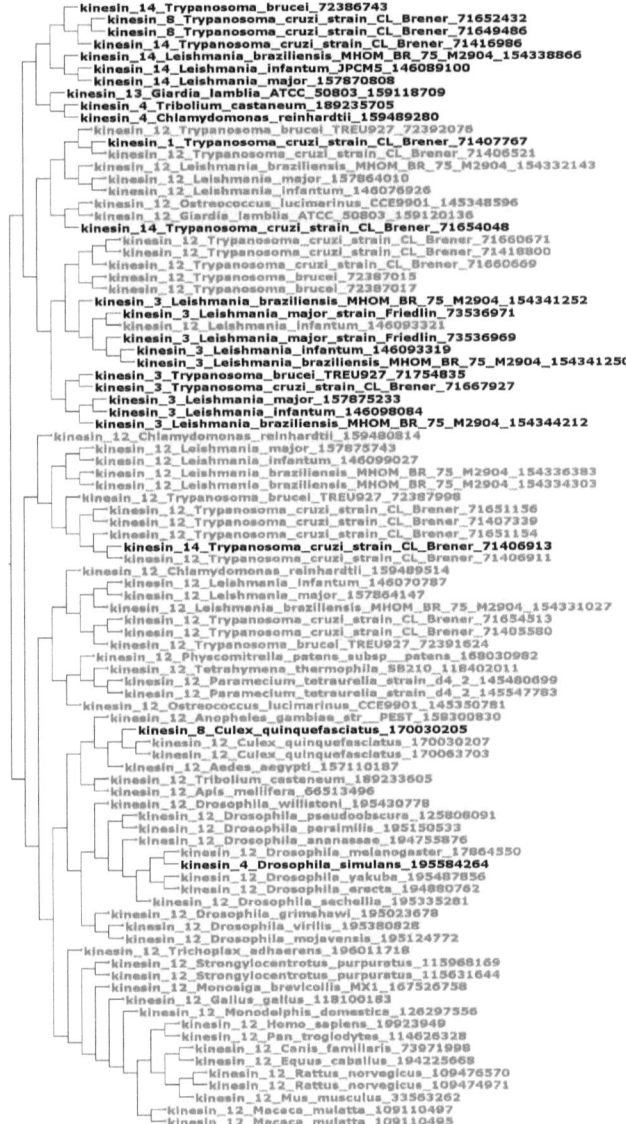

Figure 57. Sub-tree of the best-supported kinesin12 sequences. These sequences are widespread in all taxonomic groups, except fungi.

clade. All of these fit to the criteria of defining a distinct kinesin family. Thus, this clade likely represents the 14th kinesin sub-family (Figure 57). The other sequences, which do not belong to the kinesin12 in this clade, are probably false positives predicted by the CDD classification.

4.3 Bioinformatic approaches to study on kinesin-1 velocity

The significant difference in velocity between the fungal and animal kinesin-1 attracts a lot of attention from kinesin researchers. Many experiments, such as site mutagenesis and chimeric protein construction have been done in order to find out what controls the velocity. Although it has been not yet been possible to convert an animal slow kinesin into a fast one, it has been demonstrated that the velocity-control region is located inside the motor domain. Experiments have shown that many residues could be involved in controlling the velocity. However, to examining all possible combinations of residues is an unrealistic mission without the assistance of bioinformatics.

The dataset of kinesin-1 contains more than 120 sequences from 89 distinct species, which cover all kingdoms of the eukaryotes. Therefore, it is a rich dataset enough for analyzing discriminating positions between the fungal and animal sequences.

Analyses have revealed 100 discriminating positions inside the motor domain with over 80% conservation. Among these positions, 22 of them conserved in both groups, 23 are only conserved in the fungi, while the other 55 are only conserved in the animals. With the help of the high resolution crystal structures of the motor domain, two significant structural conformation changes have been determined. When mapping the discriminating positions onto the structures, many crucial combinations of residues have been exposed. These pairs seem to be important for controlling the conformation changes and binding with the microtubule. For example, position 204 of the switch I in the NcK1 motor domain contacts position 257 in the closed conformational state, preventing loop11, a main microtubule binding interface, from interacting with the microtubule. At the same time, the switch-I is fixed on the $\beta 4$ with the 203-140 binding. In the open conformational state, the structure analysis shows that the 204-257 binding is broken and loop11 is bound to the microtubule. Intriguingly, the switch-I is still connected with the $\beta 4$; however, not with the 203-140 binding any more, but with the 204-140 binding. Moreover, the discriminating residues are also responsible for the ordered structure of the loop11 in the closed conformational state; holding and releasing the β5 and many other internal structure changes while the conformation changes.

Furthermore, the chimera experiments have provided a clue about the combination of residues. This suggest that the ATP-binding pocket could be crucial for controlling the velocity and the structural interaction between the first half and the second half of the motor domain could work in cooperation with the ATP-binding pocket in order to keep the structure in the most efficient state for energy conversion.

4.4 Outlook

Bioinformatic analyses have shown great power in studying both kinesin super-family and kinesin-1 sub-families. The approaches can be easily applied to other kinesin sub-families. For example, the discriminating-position-search tool can be used for studying the differences between sub-families such as kinesin-14 and kinesin-5, which in turn could be responsible for their different functions.

The methods also can be applied to other protein families, such as myosin and dyneins. Furthermore, the evolutionary history of an entire motor protein family can be investigated.

The structural analyses of the kinesin-1 motor domain have shown the importance of the discriminating residues. Together with the conserved residues, the important structural changes from a closed to an open conformation of the motor domain have been discovered in this work. The success of this method provides a useful way to study protein structures.

Ancestral protein reconstruction is useful for understanding the evolution of a protein family. The ancestors of three fungal kinesin-1 sub-groups were successfully expressed and functionally tested. They provide new clues about the evolution of the velocity of kinesin-1. For example, the ancestor of all fungal kinesin-1s could be a slow kinesin rather than a fast one. When the ancestors of the animal kinesin-1 and the whole kinesin-1 sub-family are functionally characterized, we will have a better understanding of the evolution of the velocity of kinesin-1.

5 Appendix

5.1 Web-server

The kinesin web-server aims to provide an up-to-date kinesins dataset and many useful tools for easily analyzing kinesin sequences and structures. The web-server can be accessed under http://www.bio.uni-muenchen.de/~liu/kinesin_new/

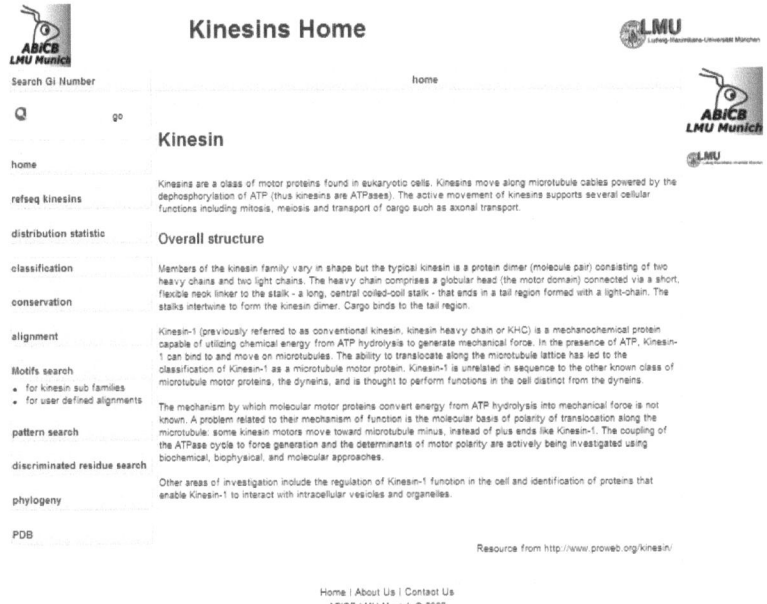

Figure 58. Homepage of the kinesin web-server

5.1.1 Classification of user defined sequence

The classification tool can be opened by clicking the "classification" link of the left navigatation menu on the web site. The user can paste protein sequence(s) in fasta format in the text area or upload a file of protein sequence(s). By clicking the submit button an automatic classification program will be run on the sequence(s).

A sample result is shown in Figure 59. CDD classification and hmmer classification results are listed.

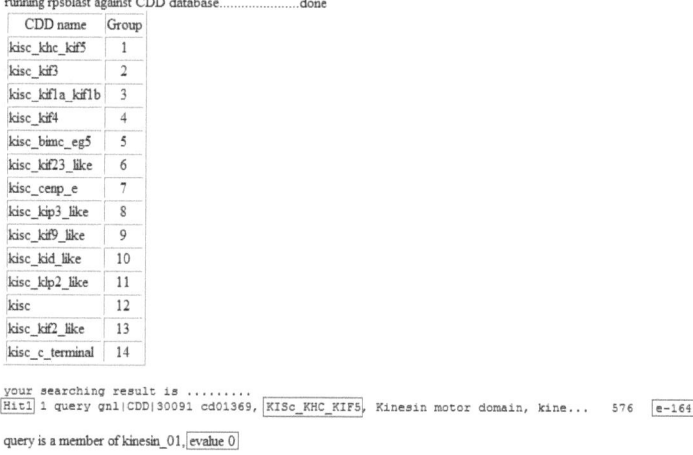

Figure 59. Classification result of gi4758650. Hit1 means the first(best) hit in the CDD database with e-value e-164 is found for the query. The corresponding class is KISc_KHC_KIF5, which represents kinesin-1. The hmmer classification result is listed in the second line. The query is clearly classified as kinesin-1 with e-value=0.

For the CDD result, CDD kinesin names and the corresponding standard kinesin names are listed in the table. Hit number is shown in the result. The smaller the number, the more significant is the classification result.

For both CDD and hmmer results, e-values are provided. If the difference between two e-values is smaller than 6, the result is considered significant. Otherwise, the one with the smaller e-value is more likely to be the correct class of the query.

5.1.2 Conservation calculation tool

The user can paste or upload his own protein alignment for calculating the conserved positions in the alignment. Three conservation methods are implemented for the calculation (see methods). Each conserved position is printed in a line with maximal 24 columns, eight columns for each method. The eight columns are method, idx, maxchar, maxcon, ref, refpos, refAA and conservation info, respectively.

Figure 60 shows a sample output of the conservation of the kinesin-1 sub-family at conservation level 1.0.

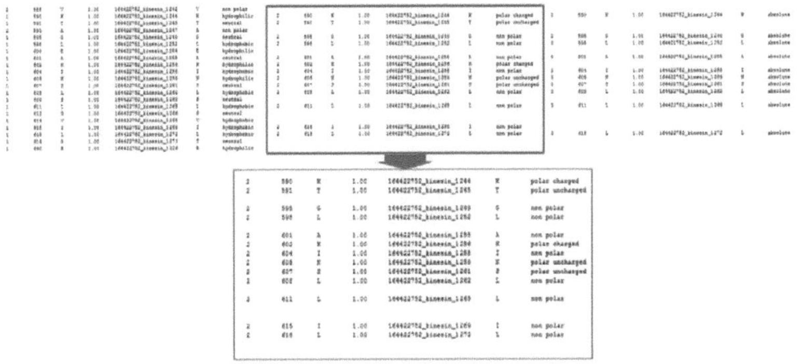

Figure 60. Sample result of the conservation calculation tool.

5.1.3 Motif search

Motifs can be searched for pre-defined kinesin groups or a user-defined alignment. Two parameters are required for the program: conservation level and gap length. Conservation level is a number between 0 and 1. It defines the minimal conservation value of the motifs to be shown. Gap length defines the maximal allowed gaps for extension of a motif when scanning the alignment. The user should define these two parameters before starting the search.

To perform the search, the user simply choose one kinesin group or uploads an alignment, defines the parameters, and clicks the submit button. A list of predicted motifs will be shown on the browser in few seconds. The user can submit an alignment of any protein set for calculation.

Figure 61 shows a sample output of the a motif scan of the kinesin-1 sub-family. The first three columns indicate the position of a motif in the alignment. Column 4 represents a reference sequence name. Columns 5-7 define the relative position of the motif in the reference sequence. Column 8 gives the average conservation score of the motif. The motif itself is listed in the last column in a simple regular expression format, where less conserved positions are displayed as small 'x's.

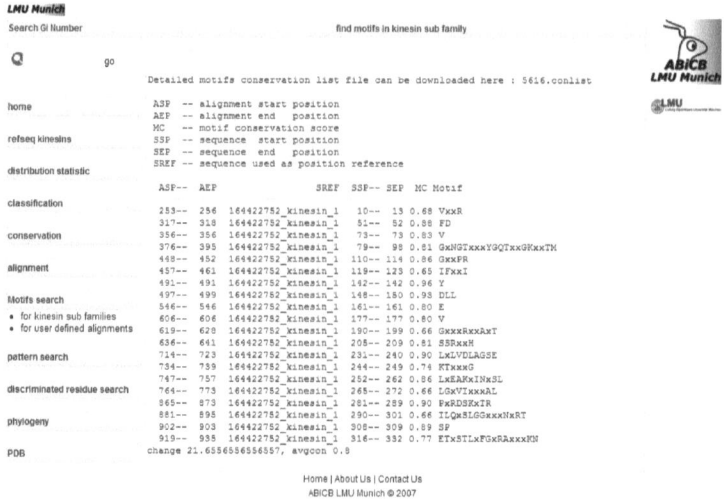

Figure 61. Sample output of a motif scan of the kinesin-1 sub-family.

5.1.4 Pattern search for known motifs

This tool is useful for searching known motifs in the kinesin database. Motifs should be written as regular expressions. Detailed descriptions and examples can be found at http://www.bio.uni-muenchen.de/~liu/kinesin_new/php/help.php#example.

An example search result is shown in Figure 62. It indicates that the search query VxxR

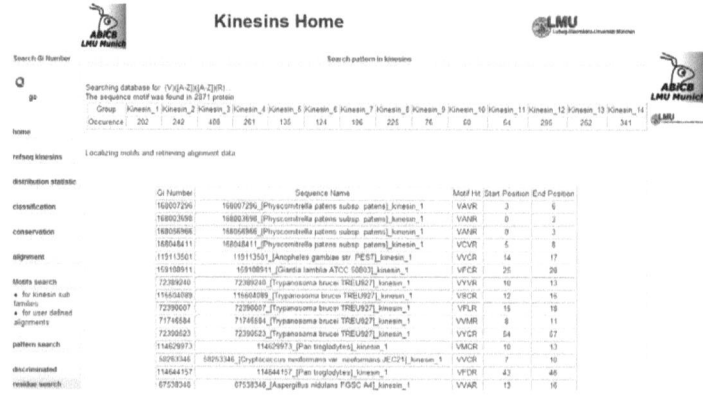

Figure 62. Sample output of a pattern search for the VxxR motif, which is a common motif of the kinesin super-family because it occurs often in each kinesin sub-family.

5.1.5 Discriminating residues search tool

Two different programs were implemented in this tool. The user can directly compare two kinesin sub-families or submit his own alignments for comparison. Before submitting a search job, a conservation level and a reference group should be defined.

Figure 63 shows the discriminating residues at conservation level of 0.8 between kinesin-1 and kinesin-14. Each line represents a discriminating residue. For each group, a reference sequence is chosen for displaying the residue and position in the alignment.

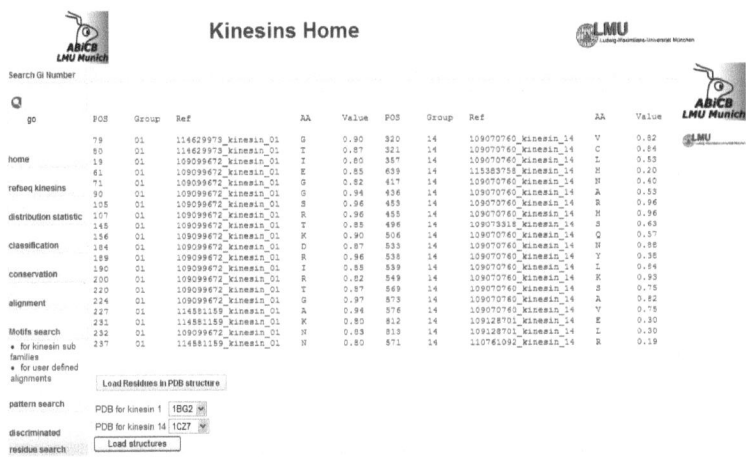

Figure 63. Discriminating residues between kinesin-1 and kinesin-14 at conservation level of 0.8.

5.1.6 Displaying discriminating residues in 3D structures

When comparing two kinesin sub-families, the available PDB 3D structures of two groups are listed under the discriminating residues list(Figure 63). The user can select one structure for each group and display all discriminating residues in the structure.

Figure 64 is a snapshot of the display of the discriminating residues between kinesin-1 and kinesin-14 (see above) in the 3D structures of 1BG2 (kinesin-1) and 1CZ7 (kinesin-14).

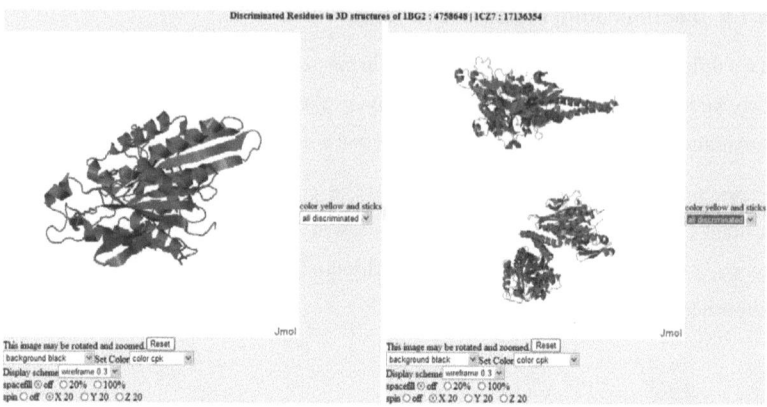

Figure 64. Comparison of discriminating residues of kinesin-1 and kinesin-14 on 3D structures 1BG2 and 1CZ7.

5.1.7 Basic information of the kinesins

Kinesin sequences can be viewed according to their kinesin sub-family classification at http://www.bio.uni-muenchen.de/~liu/kinesin_new/showRefSeqinfo.php. Clicking on a sequence name will display basic information of that sequence (Figure 65).

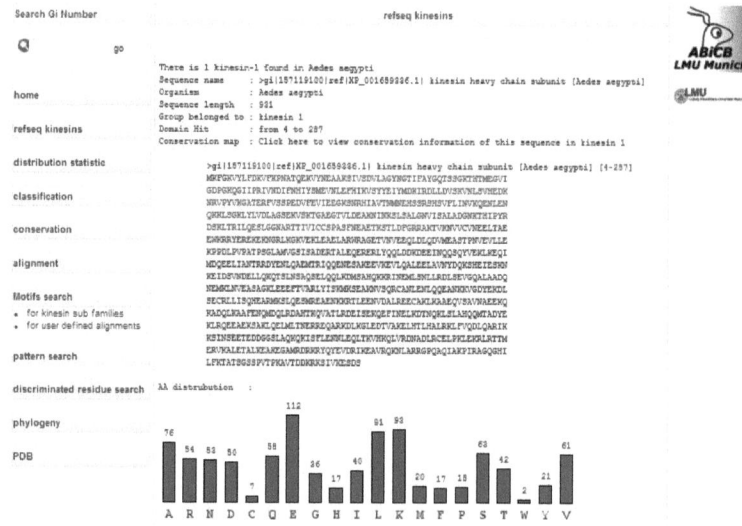

Figure 65. Example of a single sequence information page

It includes a link to the record of the sequence in the NCBI database, the protein sequence, an amino acid usage table, the predicted classification information, a highlighted blast hit and a link to the conservation of the sequence.

Three conservation scores of a sequence are calculated. Amino acids are highlighted according to their conservation scores (Figure 66).

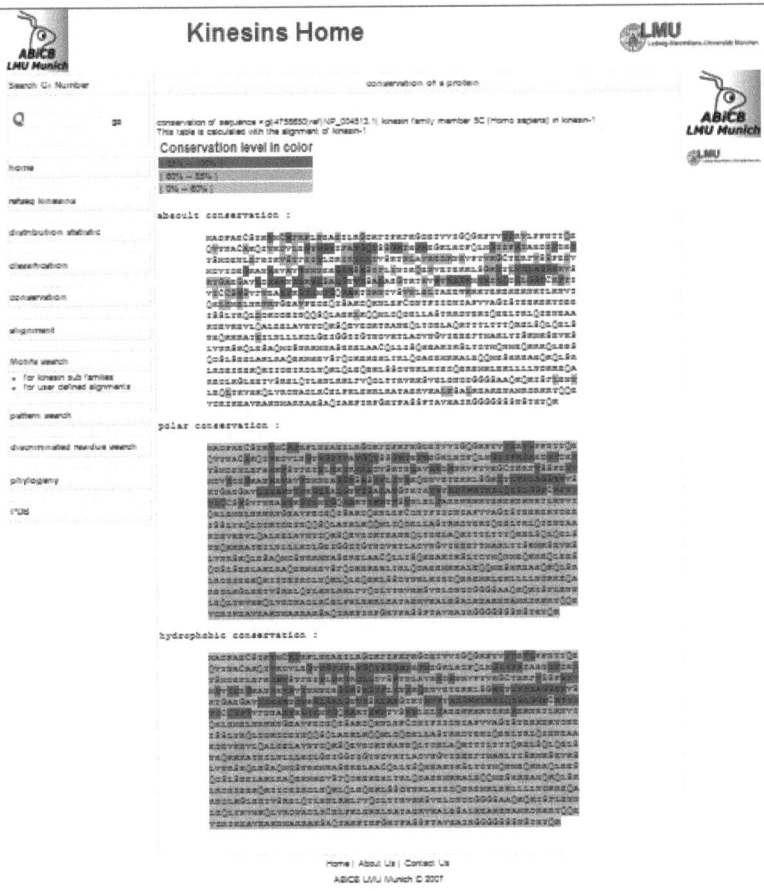

Figure 66. Example of a conservation page of a sequence.

Alignments of kinesin families can be viewed using alignment display tools. The whole alignment is displayed by default. The alignment can be downloaded via the download link. Two options can be used to modify the alignment display. One option is to highlight the conserved residues in a selected block of the alignment. The other one is to display conserved residues in a selected block of the alignment as dots, which can be useful to study differences in the sequences.

Figure 67. Alignment display options.

77 PDB structures of kinesins were classified. They can be viewed via the link http://www.bio.uni-muenchen.de/~liu/kinesin_new/showpdb.php. The display of structure is supported by Jmol [106]. Many basic display options are implemented in the website. Secondary structure elements are parsed from the PDB file, and can be displayed separately (Figure 68).

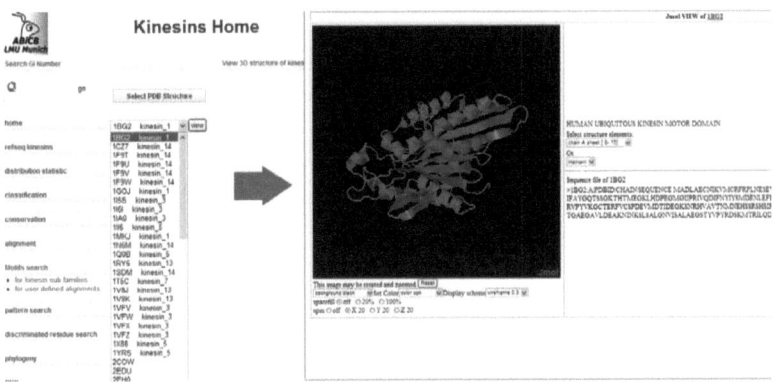

Figure 68. PDB structure display tool. The left panel is a list of all available kinesin structures. The right panel is the Jmol display of 1BG2. Basic display options are listed under the main display window. Structural elements of 1BG2 and the protein sequence are listed on the right side of the main display window.

Phylogenetic tree of 2,530 kinesin sequences is colored and labeled. Clicking on each sub-family name can display the phylogenetic tree of the individual sub-family (Figure 69).

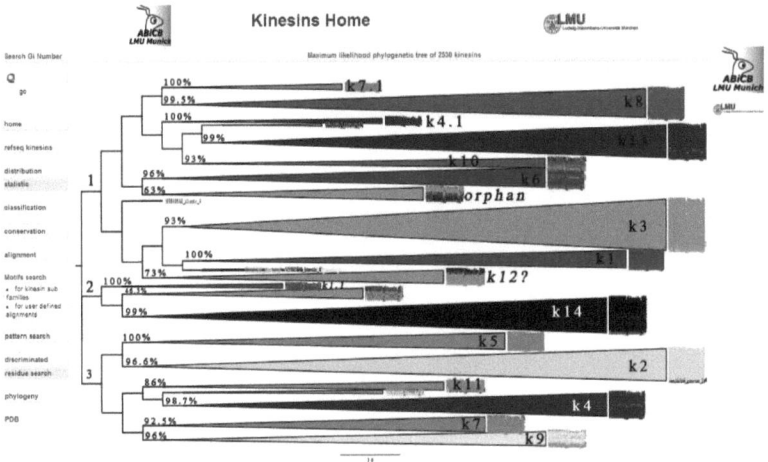

Figure 69. Phylogenetic tree of 2530 kinesins.

122

6 Bibliography

1. C. Bustamante, Y. R. Chemla, N. R. Forde, D. Izhaky. Mechanical processes in biology, Annual Review of Biochemistry, 73: 705-48. (2004)
2. Karp G. Cell and Molecular Biology: Concepts and Experiments, Fourth edition, pp. 346-358. John Wiley and Sons, Hoboken, NJ. (2005)
3. Pollard TD, Korn ED. Acanthamoeba myosin. I. Isolation from Acanthamoeba castellanii of an enzyme similar to muscle myosin. J. Biol. Chem. 248 (13): 4682-90 (1973)
4. Thompson RF, Langford GM. Myosin super-family evolutionary history. Anat Rec 268:276-89 (2002)
5. Odronitz F, Kollmar M. Drawing the tree of eukaryotic life based on the analysis of 2,269 manually annotated myosins from 328 species. Genome Biol.;8(9):R196 (2007).
6. Manfred Schliwa, ed. Molecular Motors. Wiley-VCH. (2003)
7. Lasek RJ, Brady ST. Attachment of transported vesicles to microtubules in axoplasm is facilitated by AMP-PNP Nature. Aug 15-21;316(6029):645-7 (1985)
8. Vale RD, Reese TS, Sheetz MP. Identification of a novel force-generating protein, kinesin, involved in microtubule-based motility.Cell. Aug;42(1):39-50. (1985)
9. Hirokawa N, Noda Y. Intracellular transport and kinesin super-family proteins, KIFs: structure, function, and dynamics. Physiol Rev. Jul;88(3):1089-118 (2008)
10. Walker RA, Salmon ED, Endow SA. The Drosophila claret segregation protein is a minus-end directed motor molecule. Nature. Oct 25;347(6295):780-2. (1990)
11. Vale RD, Milligan RA. The way things move: looking under the hood of molecular motor proteins. Science (journal) 288 (5463): 88-95. (2000)
12. Schnitzer MJ, Block SM. Kinesin hydrolyses one ATP per 8-nm step. Nature 388: 386-90. (1997)
13. Okada Y, Hirokawa N. A processive single-headed motor: kinesin super-family protein KIF1A. Science. Feb 19;283(5405):1152-7. (1999)
14. Lawrence C.J., Dawe R.K., Christie K.R., Cleveland D.W., Dawson S.C., Endow S.A., Goldstein L.S., Goodson H.V., Hirokawa N., Howard J., Malmberg R.L., McIntosh J.R., Miki H., Mitchison T.J., Okada Y., Reddy A.S., Saxton W.M., Schliwa M., Scholey .J.M, Vale R.D., Walczak C.E., Wordeman L. A standardized kinesin nomenclature.J. Cell Biol. 167, 19-22 (2004)
15. Kim AJ, Endow SA. A kinesin family tree. Dept of Microbiology, Duke University Medical Center, Durham, NC 27710, USA.
16. H. Miki, Y. Okada, N. Hirokawa. Analysis of the kinesin super-family: insights into structure and function. Trends Cell Biol, Vol. 15, No. 9, pp. 467-76 (2005)
17. Wickstead B, Gull K. A "holistic" kinesin phylogeny reveals new kinesin families and predicts protein functions. Mol Biol Cell. Apr;17(4):1734-43. (2006)
18. RefSeq database statistic, release 33. Jan. 16.2009
19. ftp://ftp.ncbi.nih.gov/RefSeq/release/release-statistics/archive/RefSeq-release3

3.01162009.stats.txt
20. Genebank: www.ncbi.nlm.nih.gov/Genbank/
21. PDB: www.rcsb.org/pdb/
22. NCBI: www.ncbi.nlm.nih.gov/
23. Swissport: www.swissport.com
24. PIR: http://pir.georgetown.edu/
25. EMBL: www.ebi.ac.uk/embl/
26. Altschul SF, Gish W, Miller W, Myers EW, Lipman DJ. Basic local alignment search tool. J Mol Biol 215 (3): 403-10 (1990)
27. D.W. Mount. Bioinformatics: Sequence and Genome Analysis. Cold Spring Harbor Press. (2004)
28. Zuckerkandl, E. and Pauling, L.B. Molecular disease, evolution, and genetic heterogeneity. in Kasha, M. and Pullman, B (editors). Horizons in Biochemistry. Academic Press, New York. pp. 189-225. (1962)
29. Margoliash, E. Primary Structure and Evolution Of Cytochrome C. Proc Natl Acad Sci U S A 50, 672-79. (1963)
30. Ayala, F.J. Molecular clock mirages. BioEssays 21 (1): 71-75. (1999)
31. Saitou N, Nei M. The neighbor-joining method: a new method for reconstructing phylogenetic trees. Mol Biol Evol 4 (4): 406-25. (1987)
32. Huelsenbeck JP, Ronquist F. MRBAYES: Bayesian inference of phylogenetic trees. Bioinformatics. Aug;17(8):754-5. (2001)
33. Mark Holder, Paul O.Lewis. Phylogeny estimation: traditional and bayesian approaches. Nat Rev Genet. Apr;4(4):275-84. Review. (2003)
34. Alfaro ME, Zoller S, Lutzoni F. Bayes or bootstrap? A simulation study comparing the performance of Bayesian Markov chain Monte Carlo sampling and bootstrapping in assessing phylogenetic confidence. Mol Biol Evol. Feb;20(2):255-66. (2003)
35. Pauling L. and Zuckerkandl E. Chemical paleogenetics, molecular restoration studies of extinct forms of life. Acta chemica Scandinavica 17 (89): 9-16. (1963)
36. Stackhouse J, Presnell SR, McGeehan GM, Nambiar KP, Benner SA. The ribonuclease from an extinct bovid ruminant. FEBS Lett. Mar 12;262(1):104-6. (1990)
37. Chang B.S. et al. Recreating a functional ancestral archosaur visual pigment. Molecular Biology and Evolution 19 (9): 1483-89. (2002)
38. Gaucher EA, Thomson JM, Burgan MF, Benner SA. Inferring the palaeoenvironment of ancient bacteria on the basis of resurrected proteins. Nature. Sep 18;425(6955):285-8. (2003)
39. Ivics Z, Hackett PB, Plasterk RH, Izsvák Z. Molecular reconstruction of Sleeping Beauty, a Tc1-like transposon from fish, and its transposition in human cells. Cell. Nov 14;91(4):501-10. (1997)
40. Thornton JW, Need E, Crews D. Resurrecting the ancestral steroid receptor: ancient origin of estrogen signaling. Science. Sep 19;301(5640):1714-7. (2003)
41. Chandrasekharan UM, Sanker S, Glynias MJ, Karnik SS, Husain A.Angiotensin II-forming activity in a reconstructed ancestral chymase. Science. Jan

26;271(5248):502-5. (1996)
42. Hall BG.Simple and accurate estimation of ancestral protein sequences. Proc Natl Acad Sci U S A. Apr 4;103(14):5431-6. (2006)
43. Thornton JW. Resurrecting ancient genes: experimental analysis of extinct molecules. Nat Rev Genet. May;5(5):366-75. Review. (2004)
44. Farris, J. S. Methods for computing Wagner trees. Syst. Zool . 19: 83-92. (1970)
45. Maximum likelihood: http://en.wikipedia.org/wiki/Maximum_likelihood
46. Bayesian Inference: http://en.wikipedia.org/wiki/Bayesian_inference
47. PAUP*: http://paup.csit.fsu.edu/
48. Kinesin-1: http://www.proweb.org/kinesin/FxnVessTrans.html
49. Gho M, McDonald K, Ganetzky B, Saxton WM.Effects of kinesin mutations on neuronal functions. Science. Oct 9;258(5080):313-6. (1992)
50. Hirokawa N.Organelle transport along microtubules - the role of KIFs. Trends Cell Biol. Apr;6(4):135-41 (1996)
51. Saxton WM, Hicks J, Goldstein LS, Raff EC.Kinesin heavy chain is essential for viability and neuromuscular functions in Drosophila, but mutants show no defects in mitosis. Cell. Mar 22;64(6):1093-102. (1991)
52. Berliner E, Young EC, Anderson K, Mahtani HK, Gelles J. Failure of a single-headed kinesin to track parallel to microtubule protofilaments.Nature. Feb 23;373(6516):718-21. (1995)
53. Steinberg, G., Schliwa, M. Characterization of the biophysical and motility properties of kinesin from the fungus Neurospora crassa. J. Biol. Chem. 271, 7516-21 (1996)
54. Requena, N., Alberti-Segui, C., Winzenburg, E., Horn, C., Schliwa, M., Philippsen, P., Liese, R., Fischer, R. Genetic evidence for a microtubule-destabilizing effect of conventional kinesin and analysis of its consequences for the control of nuclear distribution in Aspergillus nidulans. Mol. Microbiol. 42, 121-32 (2001)
55. Grummt, M., Pistor, S., Lottspeich, F., Schliwa, M. Cloning and functional expression of a `fast´ fungal kinesin. FEBS Lett. 427, 79-84 (1998)
56. Lehmler, C., Snetselaar, K., Steinberg, G., Schliwa, M., Kahmann, R., Bölker, M. Motor proteins and filamentous growth in Ustilago maydis. EMBO J., 16, 3464-73 (1997)
57. Schief WR, Clark RH, Crevenna AH, Howard J. Inhibition of kinesin motility by ADP and phosphate supports a hand-over-hand mechanism. Proc Natl Acad Sci U S A. Feb 3;101(5):1183-8. (2004)
58. Schäfer F., Deluca D., Majdic U., Kirchner J., Schliwa M., Moroder L., Woehlke G. A conserved tyrosine in the neck of a fungal kinesin regulates the catalytic motor core. EMBO J. 22, 450-8 (2003).
59. Holly V. Goodson, Sang Joon Kang and Sharyn A. Endow2. Molecular phylogeny of the kinesin family of microtubule motor proteins. Journal of Cell Science 107, 1875-84 (1994)
60. Kallipolitou, A., Deluca, D., Majdic, U., Lakämper, S., Cross, R., Meyhöfer, E., Moroder, L., Schliwa, M., Woehlke, G. Unusual properties of the fungal conventional kinesin neck domain from Neurospora crassa. EMBO J. 20, 6226-35

(2001)
61. Liu XI, Korde N, Jakob U, Leichert LI. CoSMoS: Conserved Sequence Motif Search in the proteome.BMC Bioinformatics. Jan 24;7:37. (2006)
62. Edwards RJ, Shields DC. GASP: Gapped Ancestral Sequence Prediction for proteins. BMC Bioinformatics. Sep 6;5:123. (2004)
63. Marchler-Bauer A, Panchenko AR, Shoemaker BA, Thiessen PA, Geer LY, and Bryant SH CDD: a database of conserved domain alignments with links to domain three-dimensional structure. Nucleic Acids Res.30:281-3. (2002)
64. Tatusov RL, Galperin MY, Natale DA, Koonin EV. The COG database: a tool for genome-scale analysis of protein functions and evolution. Nucleic Acids Res. Jan 1;28(1):33-6. (2000)
65. Kim Pruitt, Tatiana Tatusova, and Donna Maglott. The Reference Sequence (RefSeq) Project. Chapter 18. NCBI handbook. October 09, (2002)
66. RefSeq Statistic: ftp://ftp.ncbi.nih.gov/RefSeq/release/release-statistics/
67. S. F. Altschul, T. L. Madden, A. A. Schäffer, J. Zhang, Z. Zhang, W. Miller, D. J. Lipman Gapped BLAST and PSI-BLAST: a new generation of protein database search programs Nucl. Acids Res., Vol. 25, No. 17. pp. 3389-402 (1997)
68. Ben-Gal I, Shani A, Gohr A, Grau J, Arviv S, Shmilovici A, Posch S, Grosse I. Identification of Transcription Factor Binding Sites with Variable-order Bayesian Networks. Bioinformatics 21 (11): 2657-66. (2005)
69. R. Durbin, S. Eddy, A. Krogh, and G. Mitchison, Biological sequence analysis: probabilistic models of proteins and nucleic acids, Cambridge University Press, (1998).
70. Hmmer: http://hmmer.janelia.org/
71. Cox, D. R. and Hinkley, D. V; Theoretical Statistics, Chapman and Hall, (page 92) (1974)
72. LRT test online http://www.molecularevolution.org/si/resources/lrt.php
73. Larkin MA, Blackshields G, Brown NP, Chenna R, McGettigan PA, McWilliam H, Valentin F, Wallace IM, Wilm A, Lopez R, Thompson JD, Gibson TJ, Higgins DG. ClustalW2 and ClustalX version 2. Bioinformatics 23 (21): 2947-48. (2007)
74. Edgar, Robert C. MUSCLE: multiple sequence alignment with high accuracy and high throughput, Nucleic Acids Research 32(5), 1792-97. (2004)
75. J. D. Thompson , J. C. Thierry and O. Poch. RASCAL: rapid scanning and correction of multiple sequence alignments. Bioinformatics Vol. 19 no. 9 (2003)
76. CoSMoS: http://www.mcdb.lsa.umich.edu/cosmos/help.php
77. MySQL : http://www.mysql.com/
78. Stamatakis A. RAxML-VI-HPC: maximum likelihood-based phylogenetic analyses with thousands of taxa and mixed models. Bioinformatics. Nov 1;22(21):2688-90.(2006)
79. Zmasek C.M. and Eddy S.R. ATV: display and manipulation of annotated phylogenetic trees. Bioinformatics, 17, 383-4 (2001)
80. FigTree: http://tree.bio.ed.ac.uk/software/figtree/ (2009)
81. RasMol: http://www.umass.edu/microbio/rasmol/#rasmol
82. Kinesin web server: http://www.bio.uni-muenchen.de/~liu/kinesin_new/

83. Chourrout, Daniel, Philippe, Herve, Delsuc, Frederic, Brinkmann, Henner. Tunicates and not cephalochordates are the closest living relatives of vertebrates Nature. Feb 23;439(7079):965-8.(2006)
84. F. Jon Kull1,and Sharyn A. Endow. Kinesin: switch I & II and the motor mechanism. Journal of Cell Science 115, 15-23 (2002)
85. PABLO N. HESS and CLAUDIA A. DE MORAES RUSSO. An empirical test of the midpoint rooting method Volume 92 Issue 4, Pages 669 - 74 (2007)
86. Huelsenbeck JP, Ronquist F.MRBAYES: Bayesian inference of phylogenetic trees. Bioinformatics. Aug;17(8):754-5.(2001)
87. Naber N, Rice S, Matuska M, Vale RD, Cooke R, Pate E. EPR Spectroscopy Shows a Microtubule-Dependent Conformational Change in the Kinesin Switch 1 Domain Biophys J. May;84(5):3190-6 (2003)
88. Kull FJ, Endow SA.Kinesin: switch I & II and the motor mechanism Journal of Cell Science 115, 15-23 (2002)
89. Vinogradova MV, Malanina GG, Reddy VS, Reddy AS, Fletterick RJ. Structural dynamics of the microtubule binding and regulatory elements in the kinesin-like calmodulin binding protein. Journal of Structural Biology Vol.163 (2008)
90. Yamada MD, Maruta S, Yasuda S, Kondo K, Maeda H, Arata T. Conformational dynamics of loops L11 and L12 of kinesin as revealed by spin-labeling EPR. Biochem Biophys Res Commun. Dec 21;364(3):620-6 (2007)
91. Amos LA, Hirose K. A cool look at the structural changes in kinesin motor domains. J Cell Sci. Nov 15;120(Pt 22):3919-27 (2007)
92. Schultz J, Milpetz F, Bork P, Ponting CP. SMART, a simple modular architecture research tool: identification of signaling domains. Proc Natl Acad Sci U S A. 95: 5857-64.(1998)
93. Letunic I, Goodstadt L, Dickens NJ, Doerks T, Schultz J, Mott R, Ciccarelli F, Copley RR, Ponting CP, Bork P. Recent improvements to the SMART domain-based sequence annotation resource. Nucleic Acids Res. 30: 242-4. (2002)
94. Phylogenetic tree of life: http://en.wikipedia.org/wiki/Phylogenetic_tree
95. Marija Vukajlovic, Expression und Charakterisierung der evolutionären Vorläufer von Kinesin-1, (2008)
96. Song YH, Marx A, Müller J, Woehlke G, Schliwa M, Krebs A, Hoenger A, Mandelkow E. Structure of a fast kinesin: implications for ATPase mechanism and interactions with microtubules. EMBO J. Nov 15;20(22):6213-25. (2001)
97. Sack S, Müller J, Marx A, Thormählen M, Mandelkow EM, Brady ST, Mandelkow E. X-ray structure of motor and neck domains from rat brain kinesin. Biochemistry. Dec 23;36(51):16155-65. (1997)
98. Kozielski F, Sack S, Marx A, Thormählen M, Schönbrunn E, Biou V, Thompson A, Mandelkow EM, Mandelkow E. The crystal structure of dimeric kinesin and implications for microtubule-dependent motility. Cell. Dec 26;91(7):985-94.(1997)
99. Kull FJ, Sablin EP, Lau R, Fletterick RJ, Vale RD. Crystal structure of the kinesin motor domain reveals a structural similarity to myosin. Nature. Apr 11;380(6574):550-5. (1996)

100. Sindelar CV, Budny MJ, Rice S, Naber N, Fletterick R, Cooke R. Two conformations in the human kinesin power stroke defined by X-ray crystallography and EPR spectroscopy. Nat Struct Biol. Nov;9(11):844-8. (2002)
101. Sindelar CV, Downing KH. The beginning of kinesin's force-generating cycle visualized at 9-A resolution. J Cell Biol. 2007 May 7;177(3):377-85. Apr 30. (2007)
102. Baldwin, T and Lapointe, M In The Chemistry of Amino Acids: The Biology Project, (2003)
103. Gellman SH.On the role of methionine residues in the sequence-independent recognition of nonpolar protein surfaces. Biochemistry. 1991 Jul 9;30(27):6633-6. (1991)
104. Saitou N, Nei M. The neighbor-joining method: a new method for reconstructing phylogenetic trees. Mol Biol Evol 4 (4): 406-25. (1987)
105. Odronitz F, Kollmar M. Drawing the tree of eukaryotic life based on the analysis of 2,269 manually annotated myosins from 328 species. Genome Biol. 8(9):R196. (2007)
106. Jmol: http://jmol.sourceforge.net/
107. Kadane's Algo http://en.wikipedia.org/wiki/index.html?curid=10575678

6.1 Publications

108. **Liu XI**, Korde N, Jakob U, Leichert LI. CoSMoS: Conserved Sequence Motif Search in the proteome.BMC Bioinformatics. Jan 24;7:37. (2006)
109. Röhlk, C., Rohlfs, M., Leier, S., Schliwa, M., **Liu, X**., Parsch, J., Woehlke, G. Properties of the Kinesin-1 Motor DdKif3 from Dictyostelium discoideum. Eur. J. Cell Biol., (2007)
110. Joseph JM, Fey P, Ramalingam N, **Liu XI**, Rohlfs M, Noegel AA, Müller-Taubenberger A, Glöckner G, Schleicher M. The actinome of Dictyostelium discoideum in comparison to actins and actin-related proteins from other organisms. PLoS One. Jul 9;3(7):e2654. (2008)

Die VDM Verlagsservicegesellschaft sucht für wissenschaftliche Verlage abgeschlossene und herausragende

Dissertationen, Habilitationen, Diplomarbeiten, Master Theses, Magisterarbeiten usw.

für die kostenlose Publikation als Fachbuch.

Sie verfügen über eine Arbeit, die hohen inhaltlichen und formalen Ansprüchen genügt, und haben Interesse an einer honorarvergüteten Publikation?

Dann senden Sie bitte erste Informationen über sich und Ihre Arbeit per Email an *info@vdm-vsg.de*.

Sie erhalten kurzfristig unser Feedback!

VDM Verlagsservicegesellschaft mbH
Dudweiler Landstr. 99
D - 66123 Saarbrücken

Telefon +49 681 3720 174
Fax   +49 681 3720 1749

www.vdm-vsg.de

Die VDM Verlagsservicegesellschaft mbH vertritt

Printed by Books on Demand GmbH, Norderstedt / Germany